History of Computing

The *History of Computing* series publishes high-quality books which address the history of computing, with an emphasis on the 'externalist' view of this history, more accessible to a wider audience. The series examines content and history from four main quadrants: the history of relevant technologies, the history of the core science, the history of relevant business and economic developments, and the history of computing as it pertains to social history and societal developments.

Titles can span a variety of product types, including but not exclusively, themed volumes, biographies, 'profile' books (with brief biographies of a number of key people), expansions of workshop proceedings, general readers, scholarly expositions, titles used as ancillary textbooks, revivals and new editions of previous worthy titles.

These books will appeal, varyingly, to academics and students in computer science, history, mathematics, business and technology studies. Some titles will also directly appeal to professionals and practitioners of different backgrounds.

More information about this series at http://www.springer.com/series/8442

William Aspray • James W. Cortada

From Urban Legends to Political Fact-Checking

Online Scrutiny in America, 1990-2015

 Springer

William Aspray
Department of Information Science
University of Colorado Boulder
Boulder, CO, USA

James W. Cortada
Charles Babbage Institute
University of Minnesota
Minneapolis, MN, USA

ISSN 2190-6831   ISSN 2190-684X   (electronic)
History of Computing
ISBN 978-3-030-22951-1   ISBN 978-3-030-22952-8   (eBook)
https://doi.org/10.1007/978-3-030-22952-8

This Springer imprint is published by the registered company Springer Nature Switzerland AG.
The registered company address is: Gewerbestrasse 11, 6330 Cham, Switzerland

Preface

Comment is free, but facts are sacred.[1]

Many people understand that Donald Trump is interested in being the center of attention. Not only President Trump himself but also the objective viewer would say that he has succeeded with his heavy use of fake facts. As one journalist observed, "Trump showed himself to be an attentive student of disinformation and its operative principle: Reality is what you can get away with" (Rauch 2018). Many of his false pronouncements are believed by his core supporters – especially when they reinforce beliefs that the supporters already hold. But even those journalists and academics who are largely critical of Trump's fake facts provide him with satisfaction, in the spirit of all publicity (even bad publicity) being good. There are thousands of journalists and hundreds of academics who are carefully monitoring the utterances of President Trump and those of many other politicians and fact-checking these claims for their veracity. But most of these journalists and academics have two shortcomings: they do not recognize there is a long history of fake facts in America, that this Trump phenomenon is not something that only recently began and they do not understand that political fact-checking is a part of a larger phenomenon of online scrutiny that manifests itself in multiple forms.

In another book, *Fake News Nation: The Long History of Lies and Misinterpretations in America* (Rowman and Littlefield 2019), we address the first of these shortcomings. We identified eight important case studies from American history, ranging from the early nineteenth century to the near present, in which lies and misrepresentations were an important part of the unfolding of these events and their aftermaths. The events included presidential elections, assassinations, wars, business advertising, and policy debates over science and medicine.

This book continues our historical treatment by examining in detail some important developments in the period, starting in 1990 as the public Internet was about to emerge and ending around 2015 as the Trump-Clinton presidential election was

---

[1] This quotation is taken from a famous essay written on the 100th anniversary of the *Manchester Guardian* by C.P. Scott, who had served for almost 50 years as the editor of the newspaper. The article is reprinted at https://www.theguardian.com/sustainability/cp-scott-centenary-essay.

beginning to take shape. None of these historical issues were addressed in detail in the previous book.

This book also addresses the second shortcoming found in many of the existing journalistic and academic accounts of fact-checking. There is a clear evidence that some of the major players in political fact-checking – most notably snopes.com – began with a different purpose in life. snopes, and the Usenet newsgroup alt.folklore.urban from which snopes emerged, were principally about the scrutiny and possible debunking of urban legends. Some other events also occurred during this quarter century (1990–2015) that, on first glance, do not seem to have anything to do with political fact-checking: a fad for B-grade horror movies about urban legends, the popularity of truth-or-fiction television shows, the creation of a new sub-discipline of folklore studies called contemporary legends studies, and the rise of groups to educate the public about the dangers of computer viruses and other computer security risks. But we argue that these developments are all connected to one another and the connections are that they all involve determining how to act in a complex, dangerous world and that they all address this concern by adopting one kind or another of scrutiny. Thus, political fact-checking is a kind of scrutiny just as is learning how not to be duped into downloading a computer virus. These connections have scarcely been examined in the published literature. By understanding this larger domain of scrutiny, we believe we can understand better this Trumpian era of fake facts.

Who are the authors of this book, and why are we qualified to write this book? Each of us holds a doctorate in history, and each has written extensively on both the history of computing and the history of information. Each is a seasoned author, having written more than 20 books for leading academic publishers such as Princeton, Oxford, and MIT. One of us (James Cortada) is the author of *All the Facts: A History of Information in the United States Since 1870* (Oxford University Press, 2016). The other (William Aspray) was the editor in chief of the journal *Information & Culture: A Journal of History*, which is the leading journal studying the history of information. Chapter 2 (on alt.folklore.urban and snopes) and Chap. 3 (on rumors and legends surrounding the 9/11 terrorist attacks) are based on a close reading of the archives of alt.folklore.urban and snopes. The organizational histories that appear (such as alt.folklore.urban and snopes in Chap. 2, VMyths in Chap. 4, and PolitiFact and the *Washington* Post Fact Checker in Chap. 5) are stronger because of the long involvement of both authors with the study of business history, particularly when those histories involve information institutions.

How is this book organized? Chapter 1 reflects on the concept of scrutiny. Chapters 2, 3, 4 and 5 provide detailed case studies from the period 1990 to 2015. Chapter 2 examines how scrutiny happened online prior to the emergence of the public Internet and how the Internet rapidly replaced these other technologies once it became widely available. Thus, this chapter discusses not only the Usenet newsgroup alt.folklore.urban and snopes but also their predecessors such as discussion sites offered by the Internet service providers such as AOL, Prodigy, CompuServe, and The Source, as well as through electronic bulletin boards, mailing lists, and fax. Chapter 3 provides a detailed analysis of the rumors and urban legends surrounding the 9/11 terrorist attacks in 2001. This analysis enables us to examine the various types of rumors and legends, their dissemination, their debunking (when false or misleading), and their cultural meaning. Chapter 4 broadens the discussion of online scrutiny by showing how B-grade horror movies, truth-or-fiction television programming, public service activities meant to reduce the threat of computer viruses, and the rise of an academic discipline of contemporary legends all are connected to this theme of online scrutiny. Chapter 5 examines the rise of the most familiar and perhaps most important today of these efforts at online scrutiny, namely, political fact-checking. Chapter 6 briefly summarizes the material we have covered in this text and suggests opportunities for further scholarship.

We appreciate the pointers to relevant material and answers to inquiries from a number of our academic colleagues: David Bordwell on film history, Michael McDevitt on political communication, Burton St. John on advertising and public relations, and Thomas Yulsman on environmental journalism. These colleagues are, of course, not responsible for the small and large errors that we made as we tread on their turf. We also appreciate the support from the University of Colorado Boulder for William Aspray's summer research time spent on this project.

Boulder, CO, USA   William Aspray
Minneapolis, MN, USA   James W. Cortada

Reference

Rauch, Jonathan. 2018. The Constitution of Knowledge. *National Affairs* 37 (Fall). https://nationalaffairs.com/publications/detail/the-constitution-of-knowledge. Accessed 28 Dec 2018.

Contents

Chapter 1
The Concept of Scrutiny

The magic words 'on the Internet,' if inserted into nearly any sentence, seem to protect it from normal critical scrutiny.
-Nathan Myhrvold (https://www.brainyquote.com/quotes/nathan_myhrvold_541336?src=t_scrutiny)

This is a book about scrutiny in America in the quarter century leading up to the 2016 presidential election, as one way to better understand the routine use of fake facts by Donald Trump (and other politicians) and the emergence of the political fact-checking industry. While fake facts are an important part of the story, scrutiny is a broader concept that is sometimes about determining whether facts are true or false, but at other times does not particularly concern the veracity of uttered claims. Thus, checking for fake facts is sometimes the point of online scrutiny, sometimes a tool of online scrutiny, and sometimes irrelevant to the scrutiny.

The main purpose of this chapter is to provide an introduction to the concept of scrutiny, which we use in our analysis later in the book. Our major focus is about scrutiny online because of the common use of the Internet to spread lies and misrepresentations. However, we will also discuss scrutiny in other circumstances, such as scrutiny in film and television.

1.1 Online Scrutiny

Let us begin with the concept of scrutiny. It is, according to a standard dictionary definition, "the careful and detailed examination of something in order to get information about it."[1] We can get at some of the different meanings and uses of 'scrutiny' by considering various quotations about it. Sometimes, scrutiny is the means for determining the beliefs of an individual, organization, or society. As astronomer Carl Sagan notes, scrutiny is a methodological basis for the production of scientific

---

[1] https://dictionary.cambridge.org/us/dictionary/english/scrutiny

© Springer Nature Switzerland AG 2019
W. Aspray, J. W. Cortada, *From Urban Legends to Political Fact-Checking*,
History of Computing, https://doi.org/10.1007/978-3-030-22952-8_1

knowledge, and it is also important to the determination of one's philosophical or religious beliefs: "Skeptical scrutiny is the means, in both science and religion, by which deep thoughts can be winnowed from deep nonsense."[2] However, scrutiny extends beyond science and religion; it is an important attribute of the aware consumer in today's everyday life in which one is bombarded with advertising messages, as author and social media influencer Bryant McGill advocates: "Become a very cautious consumer scrutinizing everything that you allow into your mind and body."[3] Or as life coach and author Laurie Buchanan states: "Any belief worth embracing will stand up to the litmus test of scrutiny. If we have to qualify, rationalize, make exceptions for, or turn a blind eye to maintain a belief, then it may well be time to release that belief."[4]

The politics of honest election and fair governance is also a sphere in which scrutiny plays an important role, as former Senator Jacob Javits (R – NY) notes: "When scrutiny is lacking, tyranny, corruption and man's baser qualities have a better chance of entering into the public business of any government."[5] Former CIA employee and massive leaker of classified data, Edward Snowden, makes a similar point to Javits, that transparency – the ability to examine the record in detail because the details are open to outsiders – is an essential part of the political process; so that scrutiny of the documentary record of government is essential: "There can be no faith in government if our highest offices are excused from scrutiny – they should be setting the example of transparency."[6]

Another aspect of scrutiny is captured in the statement of the actress and movie producer Halle Berry: "Anytime you put a movie out it's subject to such scrutiny and such criticism."[7] The media theorist Neil Postman's report of author Ernest Hemingway's belief also speaks to this issue: "For those of you who do not know, it may be worth saying that the phrase, *crap-detecting*, originated with Ernest Hemingway who when asked if there were one quality needed, above all others, to be a good writer, replied, 'Yes, a built-in, shock-proof, crap detector.'"[8] Both of

---

[2] Brainy quotes, https://www.brainyquote.com/quotes/carl_sagan_141359?src=t_scrutiny

[3] https://www.goodreads.com/quotes/tag/scrutiny

[4] https://www.goodreads.com/quotes/tag/scrutiny

[5] https://www.brainyquote.com/quotes/jacob_k_javits_546651?src=t_scrutiny

[6] https://www.brainyquote.com/quotes/edward_snowden_551869?src=t_scrutiny

[7] https://www.brainyquote.com/quotes/halle_berry_340532?src=t_scrutiny

[8] Ernest Hemingway has famously been variously quoted and paraphrased on his comment. Raised in Michigan in the early twentieth century, exposed to the horrors of war during the First World War in Italy as an ambulance driver, living in Paris as a young writer in the 1920s, visiting battlefields during the Spanish Civil War in the 1930s, reporting for American newspapers during the Second World War, and then living in Cuba in the 1950s, Hemingway had ample opportunity to hear falsehoods and to be flooded with much misinformation. No wonder he said that everyone needed a built-in "crap detector." Neil Postman, a long-time commentator on media and other American cultural issues, heard this comment while interviewing Hemingway at the start of the 1960s, and that he repeated in well circulated speech, "Bullshit and the Art of Crap-Detection," which he delivered on November 28, 1969 at the National Convention for Teachers of English (NCTE), http://www.smirkingchimp.com/thread/8863; Kompf (2004); reported on by Postman and Weingartner (1969).

these quotations relate to the demand for authenticity in our films and other cultural artifacts such as books, plays, and dance. Authenticity is a kind of truth to self. There is an expectation that our cultural artifacts will not only square with factual information in our world, but also be internally consistent and display a kind of emotional intelligence.

Urban legends provide an interesting case in order to understand the place of scrutiny in modern society. Folklorists and sociologists created a new field of study, the study of contemporary legends, in order to understand the meaning of urban legends in contemporary American culture. These scholars noted that urban legends are sometimes true, sometimes false, and sometimes a mixture of truth and falsity; but this is not their primary concern. Instead, these scholars are interested in why urban legends are present in modern society, and explain them in part as being cautionary tales about a complex, dangerous world.

Hobbyists, such as the long-standing members of the Usenet newsgroup alt.folk-lore.urban, have also been interested in urban legends, not so much to determine the truth or falsity of particular urban legends – although they worked at doing so – but primarily to achieve two goals: to enact their believed importance of having a critical, practical mindset, together with the enjoyment derived from applying this mindset to stories that arose in an uncritical fashion on the Internet or in the press. This enjoyment – or entertainment – aspect of urban legends also showed up in various television shows such as *Fact or Fiction*, in which the audience gains entertainment value by trying to determine which of the presented stories are true and which are false. Neither the audience nor the television producers care particularly about the truth value of these stories, only about the entertainment value. This entertainment value is derived by the television viewer flexing his or her abilities to scrutinize (Fig. 1.1).

Another place where the truth or falsity of a story is secondary to another goal is with those people and organizations, such as CIAC or VMyths, that try to help the public – both individuals and institutions – to avoid falling prey to computer viruses or online mailing lists. The basic message from these organizations is that the public needs to have more online savvy, to scrutinize what they see in emails and on websites.

Scrutiny involves having a certain skepticism about messages that are received, whether in oral communication, online, or in other media broadcasts; but we do not want this skepticism to be taken too far to a point of nihilism. This skepticism must be balanced by an openness to new ideas, as Carl Sagan notes about science, but which is also true of the ordinary citizen in his or her everyday life: "At the heart of science is an essential balance between two seemingly contradictory attitudes – an openness to new ideas, no matter how bizarre or counterintuitive they may be, and the most ruthless skeptical scrutiny of all ideas, old and new. This is how deep truths are winnowed from deep nonsense."[9] Scrutiny requires thorough examination, as artist David Hockney suggests: "Photographs aren't accounts of scrutiny. The shut-

---

[9] Carl Sagan, *The Demon-Haunted World: Science as a Candle in the Dark* as quoted in Scrutiny Quotes on Goodreads, https://www.goodreads.com/quotes/tag/scrutiny

Fig. 1.1 If properly educated, Americans begin to scrutinize at a young age. (Photo courtesy of James W. Cortada)

ter is open [only] for a fraction of a second."[10] Scrutiny must also be balanced against other social goods, including privacy, as author and speaker Molly Bloom suggests: "I don't think anyone's private life stands up to public scrutiny."[11]

One problem of scrutiny is known as the *backfire effect*. This effect is increasingly evident among President Trump's hard-core supporters and similarly amongst American political far-left activist*s*. The backfire effect occurs when a message – say an accurate statement or other well-grounded effort to scrutinize some belief – has the reverse effect from what was intended, causing the recipient to dig in and embrace even more resolutely the opposite (often inaccurate) fact. Media experts, political operatives, and historians are just beginning to understand this phenomenon (Silverman 2011). However, psychologists have known about it for some time, calling it *belief perseverance*.[12] For example, if a Confederate veteran of the American Civil War believed it was better for the South that President Lincoln died – even though in the 1870s and 1880s historians began detailing how Lincoln wanted to reintegrate the Old Confederacy back into the Union in a gentle fashion – all those new facts cause the old soldier to hold faster to his belief that Lincoln was a bad person. As we showed in our earlier book on fake facts, the backfire effect occurred in every historical case we studied (Cortada and Aspray 2019). In the past

---

[10] https://www.brainyquote.com/quotes/david_hockney_470319?src=t_scrutiny

[11] https://www.brainyquote.com/quotes/molly_bloom_943563?src=t_scrutiny

[12] Baumeister et al. (2007), but known for decades, Beveridge (1950).

several decades, those who were manipulating information knew about the phenomenon and leveraged it for their purposes, helping to disseminate false information more effectively. The availability of the Internet and the concomitant extensive dumping of false and true facts has reinforced this behavior (Nyhan and Reifler 2010; Lord et al. 1979). But the main point to keep in mind is that a bias toward reinforcing earlier perceptions about an issue has long served human thinking as a general filter through which to cautiously judge alternative (newly created) "facts".

When scrutiny is avoided or its results ignored, there can be devastating consequences, as two political examples show. The first involves America's war with Vietnam. The Tonkin Resolution of 1964 gave President Lyndon Johnson the legal permission to send extra troops to Vietnam, and resulted in an additional 58,000 US military deaths and over 100,000 more wounded. Before the Tonkin Resolution, "only" 400 soldiers had died.[13] Over 2.7 million military personnel served in Vietnam., and Vietnam veterans comprised nearly 10% of their generation. Historians have concluded that the factual basis leading to the Tonkin Resolution, in response to a North Vietnamese attack on US naval forces, was largely a lie.

The second example appeared more recently, in the early 2000s, when once again a president wanted war – this time George W. Bush, who promoted the idea that weapons of mass destruction existed in Iraq and that these weapons could be used to attack other Arab oil-producing nations, NATO allies in Europe, or even Americans. Between 2003 and late 2018, nearly 4500 Americans died in the resulting war and 32,000 were wounded.[14] President Bush was called out for lying about weapons of mass destruction nearly one thousand times. United Nations inspectors went to Iraq, did not find any of these weapons, and so informed the world (Corn 2003). In this case, the fact-checkers were ignored. Historians have concluded that the premise presented to the American public for why the United States went to war with Iraq was based in false facts.

Let us next turn to the other operative word in our book subtitle: *online*. The most obvious comment to make is that the Internet provides a means with low or no cost barriers to disseminate messages that are partly or entirely lies or misrepresentations. On the other hand, the Internet also provides places such as alt.folklore.urban or snopes where people can read debunkers or even participate in the debunking process themselves. Let us look into this issue a bit more deeply by considering what we can learn about our topic from the following table (Table 1.1), which summarizes the advantages and disadvantages of online communication.

We will focus here primarily on the left-hand column, which lists the advantages of online communication, and consider how this material applies to two online scrutiny communities that we discuss in detail in Chap. 2: alt.folklore.urban and snopes. com. The right-hand column is provided mainly for better understanding of the table, as a contrast to the online communication advantages listed in the left-hand

---

[13] "Vietnam War U.S. Military Fatal Casualty Statistics," U.S. National Archives, accessed December 26, 2018, https://www.archives.gov/research/military/vietnam-war/casualty-statistics

[14] "Number of U.S. Soldiers Killed in the Iraq War from 2003 to 2018," *Statistica,* https://www.statista.com/statistics/263798/american-soldiers-killed-in-iraq/

Table 1.1 The advantages and disadvantages of online communication[a]

Advantages	Disadvantages
Flexibility: accessible 24 × 7, any place as long as you have an internet connection	**Text-based**: Predominantly relies on inputting text which can be challenging for those who don't like to write or have poor keyboard skills, but with the advance of broadband connectivity and voice and video conference technology – this will be less of an issue.
Levelling: reserved people who usually don't speak up can say as much as they like while "loud" people are just another voice and can't interrupt	**No physical cues**: without facial expressions and gestures or the ability to retract immediately there's a big risk of misunderstanding
Documented: unlike verbal conversation, online discussion is lasting and can be revisited	**Information overload**: a large volume of messages can be overwhelming and hard to follow, even stress-inducing
Encourages reflection: participants don't have to contribute until they've thought about the issue and feel ready	**Threads**: logical sequence of discussion is often broken by users not sticking to the topic (thread)
Relevance: provides a place for real life examples and experience to be exchanged	**Time lag**: even if you log on daily, 24 h can seem like a long time if you're waiting for a reply; and then the discussion could have moved on and left you behind
Choice: a quick question or comment, or a long reflective account are equally possible	**Inefficient**: it takes longer than verbal conversation and so it's hard to reply to all the points in a message, easily leaving questions unanswered
Community: over time can develop into a supportive, stimulating community which participants come to regard as the high point of their course	**Isolation**: some learners prefer to learn on their own and don't participate in the discussions
Limitless: you can never predict where the discussion will go; the unexpected often results in increased incidental learning	**Directionless**: participants used to having a teacher or instructor telling them what to do can find it a leaderless environment (and that's where tutors come in)

[a]"Advantages and Disadvantages of Online Communication", *Bang The Table*, Community Engagement Blog, 2 November 2008, last updated 9 October 2018, https://www.bangthetable.com/blog/advantages-and-disadvantages-of-online-communication-2/ (accessed 26 December 2018). The authors noted that they had drawn heavily from a blog posting on Wikipedia concerning online education, but the authors do not provide the exact citation. There is a follow-up article to this one: Crispin Butteriss, "Eight Advantages of Online Communication for Citizen Engagement," *Bang The Table*, 14 April 2010, last revised 25 October 2018, https://www.bangthetable.com/blog/eight-advantages-of-online-communication-for-citizen-engagement/ (accessed 26 December 2018). The content in our table is taken verbatim from the earlier *Bang The Table* blog posting; only the formatting is changed

column. While we do not dispute that any of these features listed in the left-hand column can be advantages of online communication, the blog post from which this chart is drawn is considerably more optimistic than we are about online practice as it is applies to scrutiny. *Flexibility* certainly is an advantage for both spreaders and debunkers of false or misleading statements. *Levelling* is also possible for both spreaders and debunkers, but it goes beyond giving an opportunity to reserved people. In fact, the Internet removes institutional barriers so that no longer does one need to be a government official, a representative of a well-financed organization, or an established media organization to participate in this conversation; this situation is strikingly different from what occurred in earlier periods in American history. The Internet widens the playing field, if not necessarily leveling it. It is certainly true that online communication generally provides a record of the communication that one can revisit; however, sometimes the things that one says online are not *documented*. It is possible that the Internet can enable reflection, but instead people may get drawn in to the rapid back-and-forth communication pattern of interactive social media, and in this environment many participants are not particularly concerned about being *reflective*. Indeed, we see considerable material online that is not reflective; and, as we argue in Chap. 2, trolling was a defense mechanism developed by the old timers on alt.folklore.urban to protect against the unreflective dissemination of urban legends by newcomers.

Relevance to one's personal life is sometimes a secondary consideration on sites such as alt.folklore.urban and snopes. The immediate task concerns evaluating the claims of some information or institution, independent of one's personal experiences. In fact, a common attribute of a successful urban legend is that it is a familiar and believable narrative communicated by somebody slightly related to the disseminator but not closely enough related in order to be checked (e.g., the disseminator's cousin's hairdresser who is located in a distant city). The relevance of urban legends comes most often not from the veracity of the facts of the narrative but instead from the legend's value as a cautionary tale in a complex, dangerous world (e.g., the dangers of letting strangers into one's home or picking up a hitchhiker – topics of two of the most common urban legends). While an individual does have *choice* about whether to provide a quick question or comment on a site such as alt.folklore.urban, there is a set of well-established rules, embodied in the site's written Frequently Asked Questions section, which are policed by the senior members of the newsgroup through social norms accompanied by swift and merciless hounding if these norms are broken. People become loyal readers of snopes or loyal participants in the alt.folklore.urban community – but only if they buy in wholesale to the social and cultural norms of the organization. alt.folklore.urban has many visitors to its newsgroup who do not, cannot, or will not adopt the social and culture norms, and most of them depart quickly. Those who do buy in to the norms often become active members of what was for many years a close-knit *community*, and this sense of community continued to be carried out not only in the newsgroup but also in private emails and in-person meetings.

Sites such as alt.folklore.urban are *limitless* for at least two reasons. Snopes is driven in large part by exogenous forces, in recent years by politicians who make claims that need to be evaluated – and what these political actors say is not under the control of snopes. Indeed, to some extent, snopes can and does pick and choose the political facts it wants to check, but its agenda is also driven about what is happening on the national and local political scenes. On alt.folklore.urban, while the site reflects the core membership's practical, rational skepticism and follows the strong social norms of participation, many of the people are there to be entertained, and the news threads often go off on tangents – some of which are flippant, unserious, or irrelevant.

What is certain is that scrutiny is widely practiced. Long before Donald Trump began running for president, it was in fashion. In 2007, journalist Craig Silverman was grousing about its popularity with respect to reporting by the media: "The level of scrutiny its critics apply to everything it does – from its fixation on the legs and ratings of Katie Couric to its serious journalistic failures – is akin to that which the press applies to public officials or corporate executives" (Silverman 2007). He stated the obvious: that the Internet made scrutiny (in particular, fact-checking) easier, that the public was doing it, not just professional fact checkers, and that readers and users of information are no longer "passive consumers of information" (Silverman 2007, 7).

References

Baumeister, R.F., et al., eds. 2007. *Encyclopedia of Social Psychology*, 109–110. Thousand Oaks: SAGE.

Beveridge, W.I.B. 1950. *The Art of Scientific Investigation*, 106. New York: W.W. Norton.

Corn, David. 2003. *The Lies of George W. Bush: Mastering the Politics of Deception*. New York: Crown.

Cortada, James, and William Aspray. 2019. *Fake Facts Nation: The Long History of Lies and Misrepresentations in America*. Lanham: Rowman & Littlefield.

Kompf, Michael. 2004. The Legacy of Neil Postman. *College Quarterly* 7 (1): 20.

Lord, C.G., L. Ross, and M.R. Lepper. 1979. Biased Assimilation and Attitude Polarization: The Effects of Prior Theories on Subsequently Considered Evidence. *Journal of Personality and Social Psychology* 37 (11): 2098–2109.

Nyhan, Brendan, and Jason Reifler. 2010. When Corrections Fail: The Persistence of Political Misperceptions. *Political Behavior* 32: 303–330.

Postman, Neil, and Charles Weingartner. 1969. *Teaching as a Subversive Activity*, 2. New York: Dell.

Silverman, Craig. 2007. *Regret the Error: How Media Mistakes Pollute the Press and Imperil Free Speech*, 6. New York: Union Square Press.

———. 2011. The Backfire Effect: More on the Press's Inability to Debunk Bad Information. *Columbia Journalism Review,* June 17. https://archives.cjr.org/behind_the_news/the_backfire_effect.php.

Chapter 2
From Debunking Urban Legends to Political Fact-Checking

> *No single truth purveyor, no matter how reliable, should be considered an infallible font of accurate information. Folks make mistakes. Or they get duped. Or they have a bad day at the fact-checking bureau. Or some days they're just being silly. To not allow for any of this is to risk stepping into a pothole the size of Lake Superior.*
> -David and Barbara Mikkelson ("False Authority" (2001) as quoted in Dean 2017).

This chapter discusses how a newsgroup (alt.folklore.urban) established to debunk urban legends evolved into a leading website (snopes.com) for political fact-checking. Part of this story concerns the evolution in information and communication technology. Today we think about online discussions being carried out on a blog or through some other kind of interactive website, in a chat room, or through mailing lists. In the early days of online life – in the 1980s and early 1990s – the media for these discussions were most commonly Usenet; discussion places established by proprietary networking service providers such as AOL, Prodigy, CompuServe, and The Source; message boards and mailing lists; and even fax. Another part of this story is the change in population of who was using the web – evolving from a narrow, highly educated group to a much larger and more heterogeneous population.

alt.folklore.urban (AFU) is a Usenet newsgroups group created in 1991. Before turning to a discussion of this particular newsgroup, we will briefly consider the broader history of Usenet. Usenet is a worldwide online discussion system formed in 1979 – thus 12 years before the public introduction of the World Wide Web. It supported thousands of discussion groups, known somewhat inaccurately as "newsgroups" (Turner et al. 2005). These newsproups were organized around particular topics such as specific hobbies or ethnic interest groups.[1] Usenet was a transitional technology. It was an advance on bulletin board systems, which were first developed only a few years earlier, in the mid-1970s, and continued to be

---

[1] For a detailed ethnographic study of a Usenet newsgroup – one concerning soap operas – see Baym (1999).

© Springer Nature Switzerland AG 2019
W. Aspray, J. W. Cortada, *From Urban Legends to Political Fact-Checking*, History of Computing, https://doi.org/10.1007/978-3-030-22952-8_2

actively used until the early 1990s. Usenet was largely made obsolete by the Internet forums and chat rooms that are much easier to use and are prevalent today.

Usenet began to decline in the late 1990s in part because of its widespread use for distributing pornography and pirated software, which both increased the amount of traffic it carried to the point that it became economically infeasible for Internet Service Providers to support and because these uses brought the Internet Service Providers under uncomfortable legal scrutiny. Today, Usenet still exists in a much-reduced form; there are many fewer Internet Service Providers that operate a news server for Usenet than there had been in the 1990s. For example, AOL stopped providing access to Usenet in 2005; and Verizon, Time Warner Cable, Sprint Nextel, and AT&T all either eliminated or reduced service, e.g. not providing service for the alt.* hierarchies, in 2008.

2.1 Usenet History and Culture

Usenet was created by graduate students Tom Truscott and James Ellis of Duke University and Steve Bellovin of the University of North Carolina at Chapel Hill, based on a technology developed at Bell Labs 2 years earlier for file transfer between UNIX-based computing systems. These systems were commonly available in universities, research laboratories, and other computer-centric organizations such as companies providing Unix products and services. Not surprisingly, the early users of Usenet were a well-educated group, many of them having degrees in computer science or engineering. To participate in Usenet required a certain amount of computer literacy: using a command-line interface rather than a more user-friendly graphical user interface as well as understanding Internet-address syntax and Unix directory structures (North 1994). Many of the early Usenet users were networking enthusiasts who did not have access to ARPANET, which was available at the time only to DARPA-supported researchers (Lee 2002).

Messages in these newsgroups were originally text based, but by the early 2000s, file transfer allowed music and image files to be embedded in messages. However, the principal purpose of these discussion groups was conversation and social interaction; and the primary medium was text (Turner et al. 2005). Messages were not sent to an individual (like an email) but instead were available to anyone signed up to the newsgroup (North 1994). Messages were sent asynchronously, threaded, and publicly archived. Individuals signed up for a particular newsgroup, which was focused on a particular topic. The thousands of newsgroups were organized into hierarchies, such as *news* for the discussion of Usenet itself or *rec* for the discussion of recreational activities. Some of these newsgroups received only one or two messages a day, while others received hundreds. Some people were very active in posting messages, while others posted only a few messages or merely observed what others were posting. Unlike the typical bulletin board, which ran on a central server maintained by a central administrator, Usenet ran in a "highly decentralized fashion that fosters bottom-up initiatives and an egalitarian culture" (Lee 2002). One user's extended metaphor of what Usenet was all about is given in Table 2.1.[2] This user was Jorn Barger, who has posted almost 10,000 messages on Usenet and who created the term "weblog".

Many of the early Usenet users were enthusiastic about the opportunities to use the medium to discuss the details of Unix and other technical topics. However, there was disagreement within the Usenet community in the 1980s about the expansion to other subjects. For example, as early as 1983 the readership of *Byte* magazine grumbled about "groupies" on Usenet who were hogging precious computing resources to talk about Star Wars and Dungeons and Dragons! (Emerson 1983) 1993 seems to have been a momentous year in the history of the Internet because of the 69% growth in a single year in the number of people connected (North 1994). By 1994, there were thriving newsgroups for astrology, and both the *alt* and *recreation* hierarchies were receiving more messages than the *comp* hierarchy, where most of the computing discussion occurred. Moreover, by this time Usenet was being used increasingly for commercial interests, especially on the *biz* and *clari* newsgroup hierarchies (North 1994).

In the early years, Usenet was enabled and to some degree controlled by a haphazardly organized collection of system administrators, spread across the world. In 1984 they formed a "Backbone Cabal", which made decisions about procedures for creating newsgroups as well as changing the hierarchies in which the newsgroups were organized – under a process called the "Great Renaming". One of the main reasons for forming the cabal was to control costs while still allowing a centralized system that could deliver messages from the mainstream newsgroups stably and instantly rather than waiting until a later time when the host computers were relatively idle. The cabal established further control by installing moderators for some newsgroups.

The control of the Backbone Cabal began to languish in 1987, only 3 years after it was formed. That year, the cabal had established a formal review and voting process for making decisions about newsgroups, which was intended to deflect criticisms that they were censors and overly controlling. Usenet members voted to create newsgroups related to sex and drugs, but the cabal refused to implement them. With the membership having a libertarian leaning, a group led by John Gilmore (who several years later co-founded the Electronic Frontier Foundation) and Brian Reid (an important computer scientist in the creation and use of markup languages) created the *alt∗* hierarchy. In contrast to the continued strict procedures for creating new newsgroups in the most mainstream parts of Usenet (the so-called Big Eight – *comp*, *humanities*, *misc*, *news*, *rec*, *sci*, *soc*, and *talk* hierarchies – which were carried by all Usenet sites), the *alt* hierarchy had loose rules for the creation of new newsgroups that did not involve a formal discussion or voting process (Lee 2002; North 1994). The *alt∗* discussion groups tended to be more specialized than those belonging to the Big Eight hierarchies. Through a grass-roots effort, alternate paths were found to transmit the messages from these more controversial alternative

---

[2] As quoted in North (1994).

Table 2.1 An insider's view of usenet

From: jorn@genesis.MCS.COM (Jorn Barger)
Newsgroups: alt.culture.usenet
Subject: Explaining Usenet to non-subscribers
Date: 24 Aug 1993 14:28:05 -0500

Someone asked me how I'd describe Usenet to someone who'd never experienced it, and all I could come up with off the top of my head was 'a big house with lots of rooms' which seemed totally lame in retrospect, so I elaborated on that with this list:

1) Each room is labelled with a topic, and if you want to discuss that topic, that room is the *one* place to do it. In this way it serves as a public utility, like a library.

2) You have no way to tell who's in a room, unless they say something! This is weird, because the Arbitron [size statistics] reports may claim 100,000 people are reading your message, but you're never sure it even got transmitted right, unless someone comments. And if you're met with silence, which you often are, you have complete freedom to interpret that silence as acceptance or rejection or whatever else you like. This really challenges one's self-doubts!

3) There's no official host to greet you, and make you feel welcome – at best, you may find a self-appointed volunteer. (This is normally the first rule of social interaction, so on Usenet you really feel the lack.)

4) Talking calmly to 100,000 or more people at once is an incredibly rare privilege, and scary as hell for almost everyone. (I'm a 'blurter' myself – sometimes my mouth just takes off and starts talking without my even thinking about the risks… but that's a real advantage here, like daring to dive into cold water…)

5) It's a lot like giving everybody who asks their own printing press and a million guaranteed subscribers! Which is a beautiful image of a democratic utopia, but brings some very serious problems.

6) These are people from all over the world, which is awesome, literally McLuhan's global village…

7) …but this includes all the global-village idiots, too! Picture trying to have a serious discussion in the middle of the mall, or the park. (Then picture the worst characters from the mall and the park *in your living room*…) I think this is the real, serious, truth-justice-and-the-American-way challenge of participatory-democracy, and it will be here that whatever solutions are found, are found first…

8) It's also like a little neutral country that shares a border with every other country, *many of whom are at war*, so it has to play host to every sort of international hostility (in the widest definition of 'nation'). Newgroups inevitably start being seen as the prize in territorial contests.

9) If someone's obnoxious, you can't do much to eject them. If you were really in the same room, everybody would just stare at them in horror, and that would be enough! Here, silent stares of horror have *no* effect.

10) When you enter the Usenet-house, you're bringing with you the mood of the room where you're reading your computer screen. So if you're at home and comfortable and totally relaxed, you can be wide open, which intensifies both the communication and the cruelties. ("Usenet, bringing the high-stress atmosphere of a city bus station right into your office and home…")

11) Flames are often the rudest sort of insult, carrying equally as much emotional pain as a slap in the face. Picture a room labelled with a topic you care about, so you're looking forward to interesting discussions and new friends, but as soon as you speak, people start coming over and slapping you repeatedly. Do you fight back, or just shut up?

12) So in this sense we're in a Wild West frontier environment, and there's a need to innovate the Usenet equivalent of a Matt Dillon…

groups (Hardy 1993). The increasing use of the TCP/IP protocol by various networks enabled increasing numbers of alternative paths around nodes controlled by the Cabal, which refused to transmit messages on behalf of these alternate newsgroups. Because of this loose control over the *alt* newsgroups, an individual *alt* newsgroup "can only die, when people stop reading it. No artificial death, only natural death".[3]

In its first 5 years, Usenet was small in its number of newsgroups and number of message postings, but beginning in the mid-1980s the membership, the number of newsgroups, and the message traffic all grew rapidly. It is hard to know the exact amount of traffic on Usenet, but one well-known communication scholar (Baym 1995) has provided a sampling that presents a clear picture of the significant growth over the decade beginning in 1984: 158 newsgroups with 4241 articles in a 2-week period of that year, compared to 5464 newsgroups and 425,320 articles in a comparable 2-week period in 1993. By 1994, Usenet was regularly accessed by more than 7 million people, and over 70,000 messages were posted each day (North 1994). Wikipedia's article about Usenet, based mostly on data from Altopia.com, shows the continued growth of Usenet: 554 K daily posts in 1998, to 1.24 M in 2001, to 5 M in 2005, to 10 M in 2008, to 20 M in 2011, to 28 M in 2017. This set of statistics is a highly inflated representation of the number of traditional news posts because it includes large numbers of automated postings, many of them spam. These statistics nevertheless provide some measure of the increased growth in traffic.

In his excellent master's thesis (North 1994), Tim North described the culture of Usenet.[4] Many of these cultural elements were true widely across Usenet, but some of the newsgroups had their own particular cultures. As a culture, these groups had rules of conduct, which were sometimes embodied in writing (typically in Frequently Asked Questions, discussed later) but often in the informal exchanges between members of the newsgroup (Burnett and Bonnici 2003). Prestige accrued in these newsgroups through the quality and quantity of postings, or by providing service to the community such as maintaining mailing lists. However, one could also capture prestige in the community simply by writing from a prestigious email address such as Stanford, IBM, or NASA.

Many of the Usenet discussion groups experienced significant conflict (Baker 2001). North observed that these conflicts were most commonly caused by one of three reasons (North 1994): ignorance of the newsgoup's norms, "malicious transgression" of the norms, or simply because of differences of opinion among members. North provided examples of specific ways in which a member might create such conflict. See Table 2.2.

This conflict was enacted in four ways, North argued: (1) "private email wars", (2) "public brawling" on Usenet, (3) complaining to a system administrator, or (4) bringing in external authorities. To try to enforce the norms of the community, Usenet used the mechanisms presented in Table 2.3.

---

[3] According to Reid, from Hardy (1993), as quoted in Lee (2002).
[4] Also see Hauben (1997).

Table 2.2 Norm violations on usenet (North 1994)

i. posting a message to an inappropriate newsgroup;
ii. posting a message to too many newsgroups [example given was posted to 40 newsgroups];
iii. asking questions that are already covered in the FAQ;
iv. posting chain letters;
v. postings that break the law of mainstream societies [example given here is of someone posing as an FBI agent];
vi. asking for unnecessary or excessive assistance;
vii. giving away plot lines (spoilers);
viii. excessively long quotations from a previous posting;
ix. posting a message from someone else's account (at best discourteous, sometimes intended maliciously to get others to flame the other person whose account you used);
x. displaying bigoted attitudes;
xi. baiting (posting flame bait);
xii. having a signature file that is excessively long;
xiii. writing in ALL CAPS; and
xiv. holding private conversations on Usenet.

Table 2.3 Usenet norms enforcement mechanisms (North 1994)

1. allowing more freedom within the alt newsgroup hierarchy;
2. appeals to system administrators or external authorities;
3. avoidance of conflict through "kill files" [using a newsreader to automatically eliminate messages sent by a particular individual or a particular newsgroup, or to eliminate particular messages that have a certain word or phrase in the title];
4. informational postings that outline expected behavior;
5. self-policing of norms by members of the Net society; and moderating of newsgroups to reduce "noise".

"Newbies" sometimes had a reluctance to post until they understood the culture. While established members of a newsgroup welcomed new members, they could respond severely if the newbies intentionally or inadvertently broke the social rules of discourse. Newbies and other members were at a disadvantage if they were not native speakers of the English language or did not have well-honed writing skills because it is what you had to say and how you said it that conferred prestige in the newsgroup.

For the most part, Usenet was eventually displaced by websites constructed to foster discussion. But some people believe that something important was lost by this change:

> The beauty of Usenet always was, and still is, it was one place you could go to and find all the topics you're interested in. You always accessed all these different newsgroups with one login, one user interface, one newsreader; and if you didn't like the newsreader you had, you went to a different one, and it was still the same articles, same place. One place you went to, whereas now they're all on webforums that are all over the place
>
> I'd have to log into one place to read stuff about aviation, and then I go to another place to read stuff that's about programming; and it's a different login, it's a different user

interface. I have to remember to go to that site to read the new stuff. It's just standard all over the net, and there's 15 different websites that have discussion boards to discuss aviation, whereas before it was one Usenet, and I really miss that. I really lament that loss, and I don't think the current generation of people using webforums and social media really understand what they lost with that.[5]

2.2   alt.folklore.urban

One of the newsgroups created under the *alt* hierarchy was alt.folklore.urban (often referred to as AFU). AFU was created at about the same time as another newsgroup (alt.folklore.computers), which was dedicated to computer folklore (Tomblin 2018). AFU was created in 1991 by Joel Furr for the collection, discussion, and debunking of urban legends. Over the 1990s, the site grew in number of members and number of postings. The 23,000 messages on AFU at the beginning grew to 71,000 messages per year by 1998. There were more than 100,000 members by 1995 – although many of them only read and did not post messages to the site (Donovan 2004; Furr 1995). While AFU membership was largely drawn from the United States, Canada, and the United Kingdom, there were some members from other English-speaking countries such as Australia and New Zealand, and also from other countries, such as Norway and Hong Kong, where English was a common second language (Richer 1998). AFU quickly became established as a principal site for the study of urban legends and also as a fact-checking site for other newsgroups on Usenet (Donovan 2004; Frentzen 1997). Donovan (Donovan 2004) has given a profile of the AFU membership[6]:

> Group members span the political spectrum, but lean leftward on social issues and issues of political conscience (reproductive rights, civil liberties, gay and lesbian rights, multiculturalism, secularism and ecumenicism) and more center-right (but with greater diversity of opinion) on matters fiscal and economic. Most rarely discuss politics or religion at all, but those who do approach those topics with ideas that are highly developed and fairly consistent.[7]

Although the term 'urban legend' had been introduced in the 1960s by Richard Dorson, director of the Folklore Institute at Indiana University, it was Jan Harold Brunvand, a professor of English at the University of Utah, who popularized the term through a series of books (Brunvand 1981, 1984, 1986, 1993, 1999, 2002,

---

[5] Tomblin (2018) (August 2). Oral history interview conducted by Alexis de Coning for this book.
[6] Some of the members of AFU included Michele Tepper, an interaction designer who wrote the first article on trolling; media scholar Clay Shirky; blogger Amanda Marcotte; the principals from snopes, Barbara and David Mikkelson; and Joel Furr, who had been one of the leaders of the alt hierarchy on Usenet.
[7] One of the Old Hats, Paul Tomblin, recalls: "I don't remember a lot of heated politics going on on AFU, although I'm pretty sure it probably did happen, because I know there were a couple of severely right-wing people who basically quit the group after we kept debunking their facts." (Tomblin 2018)

Table 2.4 Definition of *urban legend* given by AFU and snopes

An **urban legend**:
* appears mysteriously and spreads spontaneously in varying forms,
* contains elements of humor or horror (the horror often "punishes" someone who flouts society's conventions).
* makes good storytelling.
* does NOT have to be false, although most are. ULs often have a basis in fact, but it's their life after-the-fact (particularly in reference to the second and third points) that gives them particular interest.[a]
Urban legends are narratives which put our fears and concerns into the form of stories or are tales which we use to confirm the rightness of our world view. As cautionary tales they warn us against engaging in risky behaviors by pointing out what has supposedly happened to others who did what we might be tempted to try. Other legends confirm our belief that it's a big, bad world out there, one awash with crazed killers, lurking terrorists, unscrupulous companies out to make a buck at any cost, and a government that doesn't give a damn (snopes.com 2017i).

[a]alt.folklore.urban Frequently Asked Questions [Part 1 of 5], archived 7 February 1997, http://faqs. cs.uu.nl/na-dir/folklore-faq/part1.html (accessed 30 May 2017)

2004) and a weekly syndicated newspaper column he wrote on urban legends from 1987 to 1992. Prior to the Internet, Brunvand was at the center of correspondence about urban legends. He had taken both undergraduate and graduate courses from Dorson at Michigan State in the 1950s, and when Dorson moved to Indiana University to chair the Committee on Folklore in 1957, Brunvand followed him to Indiana to study for his doctorate.[8]

The term 'urban legend' refers to a fabricated story that illustrates the complexity and danger in the modern world. The word 'urban' here is a stand-in for being modern, large, and complex; in fact, urban legends are not particularly about what happens in cities. The purpose of urban legends is to provide moral guidance to others through emblematic stories, to address the anxiety that people feel in a volatile and potentially dangerous world. Table 2.4 gives the definition of urban legends provided by AFU and snopes.

One purpose of the AFU newsgroup was to provide a protected site where debunkers of urban legends could pursue their interest and receive recognition and support from like-minded people; for in the real world, debunking of urban legends was sometimes met with hostility toward the debunker, e.g. some debunkers reported getting hate mail by the believers in these urban legends. As one scholar has stated:

> Debunkers' sites such as the alt.folklore.urban archive and the *Urban Legend Reference Pages* are forms of authority on knowledge, and as such they will be challenged wherever suspicion of expertise exists. Notably, believers do not look upon the words of debunkers with relief, with the desire for the threat to be extinguished by it being 'merely an urban legend.' Debunkers were instead viewed by believers as either hostile, naive, or both. (Donovan 2002)

---

[8]Brunvand (2001) discusses both his training and how his research methods differ from those of Dorson.

Because AFU members were focused on debunking the factual basis of urban legends (i.e., that this particular incident that was reported to have happened to the storekeeper's nephew did not actually happen) but were not addressing the emotional or perhaps ethical underpinnings of the legend, they were not likely to obliterate the underlying need for urban legends. The emotional value of one urban legend could easily be replaced by that of a different urban legend. It was a secondary goal – not shared widely among AFU members – to eradicate urban legends in modern society (Donovan 2004).

Over time, however, the AFU site broadened beyond the debunking of urban legends to the skeptical discussion of other kinds of facts. As the newsgroup's FAQs state:

> The group has broadened its god-given mandate from a place for discussing urban legends (ULs) to a place for confirming or disproving beliefs and facts of all kinds, including origin of vernacular ("The whole nine yards", "Sniping like a bald giraffe"), common scientific fallacies, obscure points of history, stories of pranks, the location of Foucault's pendulums, why "Space 1999" was better than "Star Trek: TOS," "What types of salmon are there?" and so on. In other words, it's a great place to get a reality check on anything that "a friend" told you, or to compare notes about odd things.[9]

With thousands of newsgroups, it was likely that some AFU postings would be of interest not only to AFU but other newsgroups as well. Some of the most likely of these newsgroups were listed on AFU's FAQ site and are repeated here in Table 2.5. The topic of all except two of them is self-evident. Cecil Adams is the pseudonym of a syndicated question-and-answer column that started in 1973, entitled *The Straight Dope*, which would appeal to the AFU newsgroup because of the wit and extensive research it displayed in answering questions. Kibology was a humorous parody on religion created by James "Kibo" Parry in 1989 and advanced first on the newsgroups talk.bizarre and alt.slack until it received its own newsgroup; interest in kibology dissipated in the mid-1990s.

AFU had a strong culture. A competition on the FAQ page to select a motto for the group gives an inkling of the culture. Many of the nominations expressed the frustrations the old timers had with the unquestioning acceptance of newcomers ("newbies" as they called them on AFU) or the sarcasm the old timers expressed in response, e.g. "The way I heard it" or "I read it in the paper" or "My girlfriend has seen people doing this. Sources don't come more reliable." There were also farcical nominations for the official AFU mascot, the official food of AFU, hack etymologies for the word POSH, and speculations on the middle name of Captain James T. Kirk of Star Trek fame.

AFU was well known among the more heavily trafficked Usenet sites for the animosity of some of its established members (referred to sometimes on their site as Old Hats) to some of the newbies.[10] Some of these old timers felt a moral obliga-

---

[9] alt.folklore.urban Frequently Asked Questions, Part 1.

[10] The Old Hats were originally self-declared and later a more formal process of selection was used. Most of the Old Hats tended to be AFU members who had been active from the founding days. As

Table 2.5 Newsgroups with
an interest in alt.folklore.
urban content[a]

alt.folklore.computers
alt.folklore.science
alt.folklore.college
alt.folklore.ghost-stories
alt.folklore.suburban
alt.folklore.info
alt.folklore.military
alt.fan.cecil-adams
sci.skeptic
alt.tasteless
alt.usage.english
alt.flame
alt.binaries.pictures.erotica.d
alt.religion.kibology
rec.arts.startrek.misc

[a]alt.folklore.urban Frequently
Asked Questions [Part 2 of 5],
archived 7 February 1997,
http://faqs.cs.uu.nl/na-dir/
folklore-faq/part2.html
(accessed 30 May 2017)

tion to take their debunking work seriously, to be a line of defense "in an increasingly credulous world" (Donovan 2004); and they could become irritated with newbies who were there not for a serious purpose but instead to be entertained, especially if these newcomers did not follow the social norms of the newsgroup. The concern of old timers went beyond the behavior of newbies and also included their attitudes – for these newbies may have had less appreciation for the time and care that had already been expended on the study of some of the classic urban myths by the newsgroup.[11] As one of the old timers said in a post in 1993: "We love

one of them, Paul Tomblin explains: "AFU has this group we called the Old Hats, and it kind of stagnated for a while. They weren't adding new people, and so another group of people who were active in AFU who weren't becoming Old Hats formed a group they called the Young Hats. That was perfectly fine with the Old Hats, really, except there was some acrimony at one point where a couple of Young Hats became Old Hats, and we thought that when we made them Old Hats, they should resign from Young Hats. You know, we saw it as a graduation, and they didn't. So, it's like, 'But you're telling them all the stuff we're talking about in private.'" (Tomblin 2018).

[11] Donovan (2004). Many of the newbies were driven away from AFU by the trolling of the established members. However, some who made many newbie mistakes stuck it out and learned the norms and how to be a successful member of AFU. One such person was Amanda Marcotte, who wrote under the name Mandy the Mighty Mouse: "she broke just about every rule set out in the FAQ in her first month. We gave her a really hard time, and she turned around and she became a really useful and interesting contributor" (Tomblin 2018). Today, Marcotte is a well-known political blogger.

newbies here. The thing is, we like some of the newbies as friends, and we like others with ketchup."[12]

Old timers often had initial skepticism about new members who had email addresses from Internet Service Providers that served large, heterogeneous populations such as America OnLine and Compuserve because they believed it was an indicator of lack of seriousness. There was so much ill treatment toward newbies that AFU imposed Nice Week in 1998, requiring for an entire week that the old timers not ridicule the newbies. When cases of ill treatment flared up in future years, the AFU leadership threatened to re-implement Nice Week (Donovan 2004). September was a particularly dreaded time for the old timers on AFU because it was typically when the newsgroup saw a major increase in urban legends appearing online as well as an increase in new members. This was presumably because of students arriving at college and gaining access to the Internet for the first time. However, as AOL and other providers offered easy-to-use and inexpensive access to the Internet, one of the members on the Urban Legends listserv bemoaned the coming of a "permanent September" (Donovan 2004; Tomblin 2018).

AFU was a textual culture. As Old Hat Paul Tomblin noted (Tomblin 2018): "We didn't have pictures, and we didn't have videos, but we did have a lot of text, and we really prized people who could write well." In contrast to some of the chattier newsgroups, AFU hated emoticons. The attitude was:

> [Shakespeare] managed to covey emotions without emoticons. We should be able to, too, you know? …it was our intellectual snobbery. …We were smart enough to write well enough to make ourselves understood. Mostly it worked. Every now and then there were some misunderstandings because the tone did not come across well, but that was more the writer's fault than the reader's fault, we always thought.

AFU was unusual among the Usenet newsgroups for how socially close many of the active members were. There were collections to get gifts for an AFU member's wedding or help with a medical bill. Members attended each other's weddings and funerals. They held some in-person meet-ups, with an especially big one called AFU World held one time in Chicago.[13] Marriages resulted from AFU membership such as David and Barbara Mikkelson of snopes fame, and Paul Tomblin and his wife Vicky (Tomblin 2018).

The FAQs, which were managed briefly by Peter van der Linden and then for many years by Terry Chan, were an important part of the culture of AFU. On a typical day, Chan spent an hour on AFU, perhaps processing 130 messages that had arrived since the previous day.[14] The FAQs included all of the urban legends that had been thoroughly hashed out by the newsgroup. These legends were marked as T (true), Tb (true to the best of our knowledge), U (undetermined), F (false), or Fb

---

[12] Ray Depew, as quoted in Tepper (1997).

[13] Church. Bob 2018 (August 15). Oral history interview conducted by Alexis de Coning for this book.

[14] Bruckman (1994). AFU needed a place to store its material. A group of the members pitched in and they purchased a 2 GB hard drive – very large for that time – and installed it at the University of Illinois Urbana Champaign. (Church 2018)

(false to the best of our knowledge). (Richer 1998) Table 2.6 provides a sampling of the urban legends that have been hashed out by the newsgroup. During the first 10 years, approximately 400 urban legends were discussed on AFU, organized into 26 thematic categories. An example from each of the categories is presented in Table 2.6. Some of the entries are more like in-jokes than they are urban legends. Others are somewhat coarse, so the list below does not give examples from the following categories: Snuff Movies, Upstanding Legends of the Penis and Scrotum, and Hide the Salami. As one can see from the list, it is a motley collection of topics, more chosen for the curiosity and entertainment value than for any greater social value. Only a few of the sampled entries, such as the ones representing Stupid Academic Tricks, Wild Life in the Fast Lane, and Other Animal (But Non-Buggy) Crackers categories are traditional urban legends. Also note that at least one of these traditional urban legends – the one about academics – is true. Overall, the list shows the skeptical approach to the world that is shared by members of the newsgroup.

Another important element of the AFU culture was trolling, which meant "the posting of deliberately fallacious information in a particular newsgroup (alt.folklore. urban), to elicit responses from newbies who are then hounded out of the community."[15] In fact, the term *trolling*, which has become widespread online and even has migrated into everyday use in the sense of taking some action with the express purpose of provoking an angry response, apparently originated in AFU. As a word that evokes a legend, it seems appropriate for AFU. It may have been that the strong culture of in-jokes within AFU evolved into the practice of trolling (Tomblin 2018). Duncan Richer, one of the old timers on AFU, as part of a kind posting to newbies, explained how trolling worked on AFU:

> In a troll, one of the legends is changed very slightly, if at all, and then posted to the group. To distinguish it as a troll, however, the word troll must usually be put in the Subject: or Keywords: lines of the news header… Your troll succeeds when someone follows up to your message, either flaming you or putting up some "me, too" response. It is best not to reveal the troll yourself, however, leaving the job to someone big on AFU. Failing that, if you get a number of responses it is best to reveal it after a reasonable period. (Richer 1998)

Richer (1998) continues on to explain a variation on trolling, involving commonly used intentional misspellings of words in postings, to try to elicit harsh reactions from newbies. These misspellings included 'cow orker' for 'co-worker', 'doe snot' for 'does not', and 'mipslet' for 'misspelt', as well as intentionally misusing 'voracity' for 'veracity' and vice versa. As Richer remarked, "It's a mean, vicious bloodsport, but for some reason we pretty much all like it" (Richer 1998). Richer also gives specific advice for how newbies should behave to get along in the AFU newsgroup:

> To succeed on AFU, don't make your first ever post a troll. Start with some followups to someone else's thread. Don't make your own thread until you have read the FAQ back-to-front and know it properly. Gradually increase the number of postings you put up, so that you don't suddenly appear on the scene and make everyone notice. Unless you come in with impregnable sources of information, people will be instantly suspicious. After all, they were newbies once, and they remember what happened to them. Oh, yeah, and don't use unexplained acronyms, or smilies. ESPECIALLY SMILIES (Richer 1998).

---

[15] Tepper (1997); also see Burnett and Bonnici (2003).

Table 2.6 Urban legends from the AFU FAQ site[a]

Category	Rating	Urban legend
The Misappliance of Science	F	A penny falling from the height of the Empire State Building will embed in the pavement.
The 'Plane Truth (What Goes Up…)	F	Airlines use a gas to keep passengers mildly sedated and less troublesome.
DOES NOT COMPUTE	T	In 1947 a moth was found in a relay of the Harvard Mark II machine and taped into the logbook as the "first actual case of bug being found."
Twinkie Twitter	Fb	Twinkies eat mold… If mold grows on a Twinkie, the Twinkie digests it…
MAD MEDICINE	F	Hair/nails continue to grow after death.
WIND-POWERED FANTASIES	Tb	A patient's intestine exploded from cauterization during surgery due to gas.
STUPID ACADEMIC TRICKS	T	Professor lists famous unsolved problems; student thought it was homework – solves it. (The student was George Dantzig.)
STUPID PEOPLE TRICKS	T	A guy goes a-shooting at Saguaro cacti; hits one. It falls and kills him.
WHAT'S IN A WORD?	F	The expression "rule of thumb" came from an old practice that permitted husbands to beat their wives so long as it was with a stick no larger than his thumb.
HOW FIRM IS YOUR FOUNDATION?	Fb	A bridge falls down due to resonance if soldiers don't break step marching over. (Broughton suspension bridge, England, 14 April 1831, fell under soldiers' march, but probably just overloaded, not resonating.)
KILL YOUR TELEVISION!	F	Soupy Sales was canned from his television show for telling kids to send him pieces of paper with pictures of dead presidents from their parents' dresser drawers.
REEFER MADNESS	T	LSD has been sold on blotter paper with cartoon characters.
ASTONISHING ANTIPODEAN ANTICS	F	The rights to "Waltzing Matilda" are owned by an American, hence it cannot be a national anthem.
LEWD FOOD	T	Many CIA snack bars are staffed by blind people.
DISNEY DEMENTIA [AND OTHE AMUSEMENT PARK STORIES]	F	Water in the "Tunnel of Love" ride is infested with snakes.
Question authority (AND OTHER CONSPIRACIES)	F	Queen Victoria so loved *Alice in Wonderland* she requested copies of all Lewis Carroll's books. She was tricked by receipt of the copy of *Symbolic Logic*.
LEGAL BEAGLES	Tb	Gerbils are trained to sniff drugs in Canada.
WILD LIFE IN THE FAST LANE	F	As part of initiation, potential gang members drive around with headlights off at night and shoot people who flash headlights at them.

(continued)

Table 2.6 (continued)

Category	Rating	Urban legend
LEGENDS ABOUT NATATORY CAPABILITIES OF LARGE ANTHROPOID PRIMATES	T	Certain tribes of Japanese macaques sift their grain in the sea.
ASTOUNDING AVIAN ANOMALIES	F	Birds won't sit on their nests if you touch one of their eggs.
DOGGIE-STYLE and CATTY-WUMPUSS	Tb	In WW2, Russians fed dogs under tanks, then released them in battle with anti-tank explosives having an antenna-like trigger on their backs.
OTHER ANIMAL (BUT NON-BUGGY) CRACKERS	T	Alligators were once found in the sewers of New York City.
ARTHROPOD CRACKERS	F	Cactus shakes, then explodes with hundreds of scorpions/spiders.

[a]alt.folklore.urban Frequently Asked Questions [Part 3 of 5], archived 7 February 1997, http://faqs. cs.uu.nl/na-dir/folklore-faq/part3.html (accessed 30 May 2017); alt.folklore.urban Frequently Asked Questions [Part 4 of 5], archived 7 February 1997, http://faqs.cs.uu.nl/na-dir/folklore-faq/part4.html (accessed 31 May 2017)

The scholar Michele Tepper provides a cultural explanation of what happens when trolling takes place on the AFU newsgroup:

> Trolling is accepted and reinforced within the alt.folklore.urban subculture because it serves the dual purpose of enforcing community standards and of increasing community cohesion by providing a game that all of those who know the rules can play against those who do not. It works both as a game and a method of subcultural boundary demarcation because the playing pieces in this game are not plastic markers or toy money but pieces of information… Trolling, of course, is named after the fishing technique in which one lets a boat drift while dangling a line with a baited hook attached and waits to see what bites. In trolling's Usenet incarnation, the hook is baited with misinformation of a specific kind: if it is at first glance incorrect, and at second or third glance comically incorrect, in a deliberately comic way, it's probably a good troll. Only those who take the bait will post a follow-up response; those in the know will perhaps read it over a second time, chuckle to themselves, and go on… The hoped-for response to a troll is an indignant correction. It is through such a correction that the complicated play of cultural capital that constitutes trolling begins. The corrector, being outside of the community, in which trolling is practiced, believes that he is proving his superiority to the troller by catching the troller's error, but he is in fact proving his inferior command of the codes of the local subculture in which trolling is practiced. (Tepper 1997)

The old timers even marked the headers to indicate the post as being a troll, enhancing their amusement at the expense of the newbies. Successful trollers sometimes sent the incomprehensible acronymic message 'YHBT, YHL. HAND' (You have been trolled. You have lost. Have a nice day!), and they even have a name for this kind of acronymic message: 'WAFU, YN' (We're alt.folklore.urban, You're not.) (Tepper 1997). However, trolling worked as a control mechanism for the newsgroup only when the number of newbies was limited.[16] When the number of people online

---

[16] Not all of the active members of AFU were comfortable with trolling. As Paul Tomblin, a member of the Old Hats, stated: "Some guys like Ted Frank and Joel Furr just trolled the hell out of newbies, and I thought this was a bad thing, that it scared away new, interested people." (Tomblin 2018).

grew rapidly – beginning around 1993 – and the number of AFU newbies grew with it, trolling became less effective at policing the social norms. Other approaches were tried, e.g. invitation-only mailing lists (e.g. Old Hats) for those with shared normative values (Tepper 1997).

One interesting characteristic of AFU was how, increasingly over time, much of the most important conversation was conducted outside of the newsgroup, through back-channel email and a private mailing list. In the 1990s, a mailing list for the Old Hats and the New Hats was created. This mailing list became increasingly important as the AFU site became inundated with people who do not share the dedication to focus on urban legend debunking and to the appearance of many off-topic messages, including increasing amounts of spam.[17] Or as Tomblin said more succinctly: "the flood of newbies and the flood of spam" (Tomblin 2018). Another AFU member, Bob Church, attributes the diminishing importance of AFU to people moving to other lists, whether to the Old Hats and New Hats mailing lists, or to Facebook when it became active, or elsewhere (Church 2018). While the AFU site is hardly active for its original purposes today, the mailing list continues nevertheless to be active in 2019, with perhaps as many as 50 messages per day (Tomblin 2018).

Including a large number of well-educated people, the AFU newsgroup had an appreciation for research and scholarship. Its FAQ section included a bibliography of well-regarded sources for people to use in tracking down information about urban legends.[18] They also point the interested reader to the quarterly newsletter (*FOAFtale News*) and refereed journal (*Contemporary Legend*) of the International Society of Contemporary Legend Research.[19]

However, the AFU members were careful to differentiate themselves from professional folklorists. They would do research in libraries and databases, conduct experiments, and consult experts, but they did not see themselves as experts.[20] As

---

[17] Michele Tepper also pointed to the fact that there were beginning to be alternative venues for myth busting as the 1990s went on, so some people began to drift away from AFU. Tepper (2018) (August 10). Oral history interview conducted by Alexis de Coning for this book.

[18] For example, in 1997 the FAQ included the following annotated bibliographic references: Adams (1984, 1988, 1994); Bronner (1990); Brunvand (1981, 1984, 1986, 1993); Dundas (1987); Dundas and Pagter (1975); Goldstuck (1994); Klintberg (1990); Kohn (1990); Krassner (1984); Opie and Opie (1972); Poundstone (1983); Sheidlower (1995); Smith (1984); Smith and Bennett (1987, 1988); Tindall and Watson (1991, 1994); Turner (1993); Van der Linden (1989, 1991).

[19] See alt.folklore.urban Frequently Asked Questions [Part 5 of 5], archived 7 February 1997, http://faqs.cs.uu.nl/na-dir/folklore-faq/part5.html (accessed 31 May 2017).

[20] On this point of not being experts, Paul Tomblin noted: "I didn't have any particular credentials for most of the things I'd posted on AFU, except that I've got an engineering mindset, and I do a lot of reading of things. So, I would say something about, I don't know, something to do with biology, but if Diane Kelly or Ian York were to disagree with me, I certainly wouldn't argue with them, because I knew that they would have the sites to the actual journal articles that I never read. I just read the popular summaries, or whatever had shown up in the popular press. It was really great. At one point we talked about how the Old Hats should get together and create a consulting company because we have so many diverse areas of expertise.

There was one guy, Bill Nelson, who knew everything there was to know about pyrotechnics and explosives. Another guy … Well, Ian York knew everything you could think of

Terry Chan said, "we're not folklore professionals – we're hacks... We don't have any pretensions to really knowing a lot about folklore. But I think it's good to understand urban legends and have a skeptical attitude about things you hear... A lot of the knowledge we have about the outside world is actually quite frail."[21]

In fact, some of the academic scholars of folklore are skeptical about the kind of study of urban legends that AFU practiced. One of the active members of the contemporary legends society, Peter Burger, described the reaction of the well-respected but also sometimes acerbic folklorist Linda Dégh:

> dismissed on-line legend telling as an activity for the socially challenged, wondering whether 'chat-group members will eventually come to the point of leaving the safety of their homes and entering real relationships […]'. Looking at legends on the Internet, Dégh states: 'Many old legends appear in regenerated forms, but so far no continuity has captured the attention of folklorists. And without continuity and the formation of conduits, these stories succumb quickly.'[22]

Nevertheless, a few years later, Dégh argues in her book that the mass media – television, mass circulation publications, and advertising – are all important to the transmission of these legends, replacing to a considerable extent the oral tradition that folklorists had previously focused their study on (Dégh 1994; Farish 1994; Burger 2002).

2.3   snopes.com

In 1991, the World Wide Web went public. Four years later, Barbara and David Mikkelson, who were both active members of AFU, created the website snopes.com (also known as the Urban Legends Reference Pages). When they decided to create snopes, they pulled all of the material from AFU that they had contributed and asked others not to use it. As one AFU member, Bob Church, remembers:

> the big split [occurred] when David and Barbara left. It never bothered me that much. I know some people were just livid about it.
>
>   We had the FAQ, which had things I had posted about Loch Ness monster, all these true/ false things. When they left, they sent a thing saying they were pulling all the data on there that came from them and they wanted it off the FAQ and they didn't want anyone using it.

---

about prion disease and viruses and that sort of thing. I mean, that was his area of research. Diane Kelly, well, actually her main field of research was on animal penises, which came up more often than you would think, mostly because we were trying to get Diane to say things about them. There were guys like Angus Johnson, who had done a lot of research about union movements in the 1930s and '40s, and that sort of social science, social and political stuff. He was a great resource on that sort of thing, 'cause like I said, a lot of us were engineers. We didn't know, I didn't know anything about political stuff in the 1930s back then. (Tomblin 2018)

[21] Chan, as quoted in Bruckman (1994).

[22] Burger (2002), including Burger's quotation of Degh (2001).

> Which from a business standpoint, that's the way they had to do it. But it did not go over well, that's the thing.
>
> Not so much the specific things they took, but it was just that we were all doing this together. We post these things and the database gets bigger, and then they took theirs. It wasn't like anyone was saying, "Wait, no, I'm the one researched this one."
>
> We're all this club, and then they aren't. And they not only aren't, they take their stuff with them. (Church 2018)

snopes provided a way to make what was accessible through AFU available to a larger public because the technical and social barriers of getting onto and navigating Usenet were largely eliminated by using a website as the medium. Snopes has become the most important website not only for tracking urban legends, but also a leading site for online fact-checking, especially political fact-checking (snopes.com 2017a). The site has adopted a broad definition of urban legends: "We employ the more expansive popular (if academically inaccurate) use of 'urban legend' as a term that embraces not only urban legends but also common fallacies, misinformation, old wives' tales, strange news stories, rumors, celebrity gossip, and similar items."[23] The Mikkelsons also moderated a list entitled The Urban Legends Digest, which in the late 1990s had more than 500 members (Donovan 2004).

David Mikkelson (born c. 1960) grew up in the San Fernando Valley in southern California and studied computer science at the University of Texas at San Antonio. He has taken numerous non-degree courses since then at California State University, Northridge. He has worked as a software engineer for several companies. While working for Digital Equipment Corporation, he became fascinated with the company's internal message board system, which not only held company product information but also information about hobbies. This led him to join both AFU and another Usenet newsgroup, rec.arts.disney. Both of these groups were discussing a mysterious private club at Disneyland. His curiosity was piqued, he placed an ad in the *Los Angeles Times* for information about Club 33, and a member contacted him and invited him to visit (Walker 2016). This reinforced his curiosity and interest in doing research to get to the bottom of things that interested him. He became an active member of AFU (Blank and Howard 2013) and gained a reputation for being the newsgroup's foremost troller, under his online name snopes:

> snopes is AFU's trollster extraordinaire. He has been known to incite violence on a regular basis, especially in any thread involving Canada. Do not believe any message posted by snopes, in fact assume whatever he says is false unless you have direct proof to the contrary. snopes is to be avoided at all costs. Do not even make the attempt to flame him in a thread he has started. This will only make you troll-bait, no matter what you have to say. Basically, the nature of AFU has put snopes in a no-lose situation, and you can't do a thing about it. Just follow his trail of devastation as newbies fall hook, line and sinker for his obvious falsehoods and post in to correct. It can be quite fun to watch. (Richer 1998)

David Mikkelson chose the name snopes for his AFU messages, and later for the website, based on a trilogy of novels by William Faulkner: *The Hamlet*, *The Town*, and *The Mansion* (Faulkner 1940, 1957, 1959). It was the name snopes itself rather

---

[23] snopes.com, Frequently Asked Questions.

than the values embodied in the poor Mississippi Snopes family that attracted him, inasmuch as the family included a murderer, a pedophile, a mentally disabled person with zoophilia, a bigamist, a voyeur, a pornographer, a thief, and a venal politician.

Barbara Hamel was in her thirties, living in Ottawa, married, working as a secretary and a bookkeeper when she was attracted to AFU. She had plans to become a journalist and applied to Ryerson University in Toronto. However, when she came down with Crohn's disease, she abandoned those plans (Dean 2017). She met David through AFU; and on their first date when she came to California to visit him, he took her to the UCLA library to do research on urban legends. She moved to California to be with David in 1994, and they were married in 1996. They ran the snopes website as a hobby from their small home in the suburbs of Los Angeles, serving as webmasters and writing all of the articles themselves. David continued to work as a software engineer, while Barbara spoke of herself as "just a housewife" (Seipp 2004).

Although the website was run at first as a hobby, the quality was sufficiently high that the leading urban legend scholar, Jan Harold Brunvand, was sufficiently impressed that he did not feel the need to start his own website. There was some division of labor in the stories they researched, based on personal interests. David focused on Coca Cola, the Beatles, Disney, and sports; while Barbara focused on business, politics, horror, and crime (Hochman 2009). David continued to take courses at California State University, Northridge, and this enabled the Mikkelsons to use the university library and its online databases to carry out their urban legend research. To give more *gravitas* to written requests for information when conducting their research, they formed the San Fernando Urban Legends Society and had stationery printed up for this fictional society (Richer 1998; Walker 2016) (Fig. 2.1).

David Mikkelson explained that there are a variety of reasons why people pass along urban legends – some involving social good, others for less beneficial reasons:

> A lot of it is just people's desire to do good. They think they're being helpful by passing along something – a piece about a missing child, or warning you about some sort of crime that you might fall victim to. Some of it is just people looking to show off. 'I'm smart. I know this and you don't.' Some of it is attempts to prove other people wrong about things, usually of a political nature. A lot of things that are truly urban legends: things that have narratives, that have plots, and morals. They're often a way that people encapsulate and pass along fears and anxieties. A lot of what we see is directly related to what's going on in the world. They're also tacit ways of expressing or reinforcing prejudices. Maybe a crime rumor that has to do with gangs or Mexicans. And it's – 'Well, I'm not saying this about whatever group. It's the story I heard says that they're doing whatever.' It's sort of a camouflage." "…you see a pattern in these kind of things: that when a new technology comes out, you start seeing disaster stories about that technology. And there is an underlying principle there. That people are wary of new technologies till they've been established and proved safe.[24]

---

[24] As quoted in Pogue, David (2010).

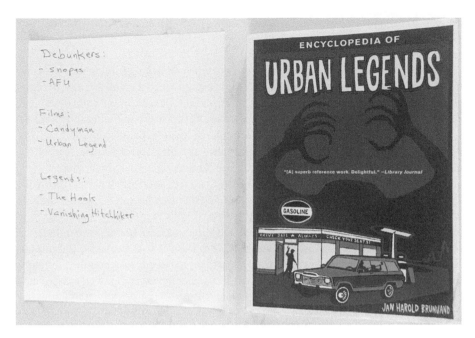

Fig. 2.1   Desk of an urban legends scholar. (Photo courtesy of William Aspray)

Traffic on the snopes website increased substantially as a result of the controversies over the 2000 U.S. presidential election between George W. Bush and Al Gore, and the 2001 9/11 terrorist attacks (Nissen 2001; Stelter 2010; Walker 2016). One author credits the 9/11 terrorist attacks as the inflection point around which snopes stopped being a hobbyist site to debunk curiosities and became a major fact-checking organization (Dean 2017). In 2002, the website was receiving 40,000–60,000 visits per day (Bond 2002). In 2002, David quit his software engineering job to work full-time for snopes, with advertising revenue from the website making this possible. Hurricane Katrina in 2005 and the U.S. presidential election of 2008 (especially the rumors circulating about Barack Obama's foreign origins) led to similar increases in traffic on the snopes website (Walker 2016). There was also a big uptick in snopes web traffic during the 2016 U.S. presidential election (Streitfeld 2016).

By 2009, snopes was a sufficiently important site in politics that it was being scrutinized by conservatives for alleged liberal bias. FactCheck.org, run by the widely-respected Annenberg Public Policy Institute, investigated these claims and found that snopes was not being funded secretly by billionaire George Soros but instead received all of its funding through advertising on its website. It also found that David was a Republican turned Independent, while Barbara was a Canadian citizen who could not vote in U.S. elections.[25] By 2010, the site was receiving 300,000 visits a day, and the Mikkelsons had hired two people to help them with the mail. (Pogue 2010)

In 2014, Barbara stopped working on the snopes website, citing health reasons. Not long thereafter, she moved to Las Vegas and divorced David in a highly contested legal battle.[26] She no longer holds an economic interest in the website. David could not continue to operate snopes in the way that it had been without Barbara's involvement, even with the help of a couple of people to answer mail; so, he professionalized the operation. Snopes moved into office space, hired a geographically dispersed team of about a dozen people to help not only with answering mail but also with writing and editing, and entered into a partnership with a San Diego firm, Proper Media, which improved the website and helped Mikkelson sell advertising (Streitfeld 2016; Walker 2016).

Increasingly since 2010, traffic to the site was not so much about urban legends but instead about political fact-checking. One reason is that extensive debunking online has mostly eliminated some of the classic urban legends such as the alligators in the New York City sewers.[27] As political fact-checking on the snopes website has grown, David Mikkelson has taken greater care to appear politically neutral, not that this stops his critics. He carefully avoids making partisan political comments, and snopes bans advertising that is political in nature. Political fact-checking has a different rhythm and time frame from debunking urban legends. Political stories tend to appear suddenly and fade rapidly. For snopes to have a significant voice in this arena, it must address these claims rapidly and use its web site and social networking to get the word out about its findings (Walker 2016; Stelter 2010).

Confirmation bias has always played a role on the Internet; people often believe stories, whether they are true or not, if they confirm the reader's beliefs. (Walker 2016) What is different today is that the financial or political rewards are high and the penalties negligible for publishing news stories without adequately checking them out or even publishing stories known to be false. This kind of fake news has become common, especially since the 2016 U.S. presidential election (Streitfeld 2016).

One famous case is that of departed.com, which was created by Beqa Latsabizde from the Republic of Georgia during the 2016 U.S. presidential election campaign, to feed a steady stream of pro-Trump, anti-Clinton stories onto the web. These stories were a combination of fact and fiction, typically cut and pasted from other stories found online. Departed.com earns a few thousand dollars a month from Google, which pays pennies each time a reader clicks on an ad embedded in the departed.com website (Higgins et al. 2016; Isaac 2016; Streitfeld 2016). Google has been upset about its enabling role in the spread of this fake news; and Mark Zuckerberg, the CEO of Facebook, is also sensitive to criticisms that fake news on

---

[25] Walker (2016); also see Fader (2012); Novak (2009).

[26] By this time, David was already dating Elyssa Young, whom he later married. The tabloids had a field day with this story because Young had a background as a sex worker. She works at snopes as an administrative assistant. (Dean 2017)

[27] Jan Harold Brunvand as cited in Hochman (2009).

his social networking site may have affected the outcome of the presidential election. Both companies are taking steps to make it harder for their sites to be used for these purposes. Google has set a policy of banning websites that spread fake news from using its online advertising service. Facebook is placing labels on websites where stories had been shown by a reputable group to be false and taking other means to make it harder for other organizations to gain financially from the spread of fake news online. Facebook hired snopes as well as several other organizations to do fact-checking for them (Streitfeld 2016; Isaac 2016).

As of this writing, snopes is ranked 593 in the United States and 2399 worldwide in the number of visitors to its website.[28] However, the site received substantially heavier traffic during the 2016 U.S. presidential election, when it was ranked in the top 1500 websites worldwide. In this usage snapshot, 72% of the visitors were from the US; together, Canada, the UK, and Australia represented almost 10% of the visitors. The next most frequent country of origin was Japan, at 1.3%. Compared to the average website, visitors to snopes were more likely to be men, more likely to have a college education, much less likely to have no college education, more likely to have accessed the site from home, and much less likely to have accessed the site from school.

In this era in which websites are often charged with political bias, snopes has been assiduous on its website in providing transparency about the ways in which it conducts its business.[29] The names of all the professional and support staff are listed on the site, and the website provides a means for contacting any of them. All of the snopes' employees are required to complete a course on fact-checking offered by the Poynter News University, the leading center for online education in journalism; and the staff is prohibited from "donating to, or participating in, political campaigns, political parties, or political advocacy organizations" (snopes.com 2017b).

Their lengthy offering of transparency begins with a discussion of the step-by-step methodology by which the staff addresses a particular urban legend or possible political fact, including the use of multiple non-partisan sources (with peer-reviewed journals and government agency statistics especially highly valued) and the checking of their findings by multiple members of staff (snopes.com 2017d). At least some educators approve of the snopes process and use the site as a model for critical and analytical evaluation of web sites and stories found online.[30]

The website also discusses the way in which snopes decides which issues to write about:

> ...we have long observed the principle that we write about whatever items the greatest number of readers are asking about or searching for at any given time, without any partisan considerations. We don't choose (or exclude) items for coverage based on whether they deal with Republican/Democratic, conservative/liberal, or religious/secular issues. We also don't impose our own judgments about whether a given item's perceived importance, controversiality, obviousness, or superficiality (or lack thereof) merit our addressing it...

---

[28] Alexa, statistics from 28 May 2018. http://www.alexa.com/siteinfo/snopes.com

[29] For an example of a site claiming snopes has political bias, see goodnature (2016). There are many more!

[30] See DeGroot (2011).

Table 2.7 snopes rating system (snopes.com 2017f)

true
mostly true
mixture
mostly false
false
unproven ["This rating indicates that insufficient evidence exists to establish the given claim as true, but the claim cannot be definitively proved false."]
outdated ["subsequent events have rendered their original truth rating irrelevant (e.g., a condition that was the subject of protest has been rectified, or the passage of a controversial law has since been repealed)."]
correct attribution
misattribution
legend ["This rating is most commonly associated with items that are "pure" urban legends: events so general that they could have happened to someone, somewhere, at some time, and are therefore essentially unprovable."]

> The inputs we use for the process of determining reader interest include the tabulation of terms entered into our search engine, reader e-mail submissions, comments and items posted to our Twitter and Facebook accounts, external social media posts, Google Trends, Twitter's Trending Now, Facebook's Trending Topics, and items flagged for review by Facebook users as part of our partnership with Facebook. (snopes.com 2017h)

Because the range of topics addressed by snopes extends beyond urban legends to include "e.g., trivia, rumors, hoaxes, common misconceptions, odd facts", the ratings system is slightly more extensive and flexible than the one used by AFU. See Table 2.7.

The transparency discussion then turns to a more extensive consideration of sources than was addressed in the methodology section described above. Snopes lists all of its sources on the website. It attempts to contact the source of the claim, as well as topical experts, for further information as a regular feature of the research process. It also consults printed sources that it deems to be reliable and non-partisan. Moreover, to educate its readers, snopes informs "readers that information and data from sources such as political advocacy organizations and partisan think tanks should be regarded with skepticism"(snopes.com 2017g).

With regard to correction policy, snopes notes that it tries to promptly correct "errors of fact and to clarify any potentially confusing or ambiguous statements in our articles." Readers are provided with a way in which to offer corrections and other feedback. When corrections have been made to an entry on the website, this change is noted in a clear way (snopes.com 2017c).

The website also provides transparency about the organization's sources of income and what these funds are spent on. The site points out that there are no hidden owners or partners and that revenue is virtually all from advertising on the website; and a summary breakdown of expenditures for calendar year 2016 is provided: 34% for employee compensation; 37% for technical support, development, and advertising; 12% for overhead, travel, education, and training; and 17% carried over for future expansion and expenses (snopes.com 2017e).

The transparency theme is continued in the Frequently Asked Questions section of the snopes website, where one question notes that sometimes snopes is writing about a topic that is the subject of one of its advertisements and wonders whether snopes was being paid to write about this topic. Snopes replies:

> Some of the advertising carried on our site is supplied by Google's AdSense program, a system that scans the text of web pages and automatically displays ads for products and services related to keywords appearing on those pages. We have no control over which ads Google chooses to display on any given page, nor do we have any business relationships with those advertisers. Also, since we have a large and diverse variety of advertisements rotating through our site every day, and we cover a wide range of topics on our site, occasionally an advertisement for a particular business or product may display on a page that includes editorial content about that same business or product out of sheer coincidence. We are not (and never have been) paid or provided with any other form of remuneration in exchange for writing about a particular topic.[31]

Nevertheless, one serious information scholar, Kalev Leetaru, has been critical of snopes's methods and transparency (Leetaru 2016). Leetaru's article was written to try to determine the veracity of a salacious article that had appeared in the British newspaper, *The Daily Mail*, the day before (Goodman 2016). Claiming to have access to the legal records associated with the contentious divorce between Barbara and David Mikkelson, *The Daily Mail* argued that David had embezzled funds from snopes to pay for prostitutes and other personal indulgences, and that one of these prostitutes had become his second wife and was placed on the snopes' payroll. To add insult to injury, the new wife had run for political office (a Libertarian Party candidate for Congress in Hawaii in 2004),[32] breaking the snopes' rule about not employing people who had partisan political affiliations. Leetaru entered into communication with David Mikkelson, who was circumspect in answering questions, claiming that he was bound by the confidentiality clauses in the divorce settlement not to address some of the issues raised by Leetaru. In fact, Leetaru never was able to determine the veracity of the claims in *The Daily Mail* article – and we have no intention to try to do so either. But in the course of the communication with David Mikkelson, Leetaru found the snopes' methodology wanting. See Table 2.8 for Leetaru's criticisms.

Leetaru concluded from this exchange that it presents "a deeply troubling picture of a secretive black box that acts as ultimate arbitrator of truth, yet reveals little of its inner workings." A more charitable conclusion might be that snopes has thrived for many years by a formal, non-academic but conscientious process, which might not hold up to academic scrutiny but which may be practically sound.

---

[31] snopes.com, Frequently Asked Questions.

[32] See Dean (2017).

Table 2.8 Leetaru's criticisms of the snopes methodology (Leetaru 2016)

1. When asked about whether snopes was employing someone who had expressed partisan political leanings by having run for political office, Mikkelson responded indirectly, saying "It's pretty much a given that anyone who has ever run for (or held) a political office did so under some form of party affiliation and said something critical about their opponent(s) and/or other politicians at some point. Does that mean anyone who has ever run for office is manifestly unsuited to be associated with a fact-checking endeavor, in any capacity?"
2. When asked for the detailed criteria used to hire fact checkers, "David demurred, saying only that the site looks for applicants across all fields and skills. He specifically did not provide any detail of any kind regarding the screening process and how Snopes evaluates potential hires. David also did not respond to further emails asking whether, as part of the screening process, Snopes has applicants fact check a set of articles to evaluate their reasoning and research skills and to gain insight into their thinking process."
3. When asked how snopes assesses its fact checkers and in particular conducts intra- and inter-reliability assessments, Mikkelson did not respond.
4. When asked why snopes only lists a single person as the author of a particular entry on its website rather than listing all of those who contributed to writing and vetting the entry, Mikkelson simply replied that it was their practice to simply credit the initial author. As Leetaru argues, "Not only does this rob those individuals of credit, but perhaps most critically, it makes it impossible for outside entities to audit who is contributing to what fact check and to ensure that fact checkers who self-identify as strongly supportive or against particular topics are not assigned to fact check those topics to prevent the appearance of conflicts of interest or bias."
5. Leetaru also did not receive a response from Mikkelson about (a) "why Snopes fact checks rarely mention that they reached out to the authors of the article being fact checked to get their side of the story"; (b) "why the site does not have a dedicated appeals page for authors of stories which Snopes has labeled false to contest that label"; or (c) "whether Snopes has a written formal appeals process or how it handles such requests." To this, Leetaru concluded that Snopes carries out its "fact checking from afar."

snopes experienced another kind of crisis in 2017. As David Mikkelson explains:

> Snopes.com, which began as a small one-person effort in 1994 and has since become one of the Internet's oldest and most popular fact-checking sites, is in danger of closing its doors. So, for the first time in our history, we are turning to you, our readership, for help.
>
> Since our inception, we have always been a self-sustaining site that provides a free service to the online world: we've had no sponsors, no outside investors or funding, and no source of revenue other than that provided by online advertising. Unfortunately, we have been cut off from our historic source of advertising income.
>
> We had previously contracted with an outside vendor to provide certain services for snopes.com. That contractual relationship ended earlier this year, but the vendor will not acknowledge the change in contractual status and continues to essentially hold the snopes.com web site hostage. Although we maintain editorial control (for now), the vendor will not relinquish the site's hosting to our control, so we cannot modify the site, develop it, or – most crucially – place advertising on it. The vendor continues to insert their own ads and has been withholding the advertising revenue from us.
>
> Our legal team is fighting hard for us, but, having been cut off from all revenue, we are facing the prospect of having no financial means to continue operating the site and paying our staff (not to mention covering our legal fees) in the meanwhile.[33]

When the Mikkelsons divorced, Barbara sold for $3.6 M her half interest in snopes (or more correctly, her half interest in Bardav, the parent company for snopes, which

---

[33] https://www.gofundme.com/savesnopes, accessed 27 July 2017.

the two Mikkelsons had created in 2003) to five individuals associated with the vendor mentioned above, Proper Media, that provided services to snopes.[34] David Mikkelson terminated the agreement with Proper Media in 2017, and this led to a legal battle between them. Proper Media cut off revenues to David Mikkelson.[35] He responded by launching an online donation campaign through gofundme, with a goal of $500,000. In only 3 days the campaign had collected over $644,000 – giving some indication of the widespread interest in the service provided by snopes. Whether this will enable Mikkelson to regain control and redevelop a reliable advertising revenue source is not clear at this time.

2.4   Conclusions

This chapter has focused on recounting the origins and history of snopes, one of the leading fact-checking sites on the web. The origins of snopes in the Usenet newsgroup AFU has a bearing on how snopes conducts its business. The most active participants in AFU came from many walks of life (although there was an overrepresentation from the tech community, especially in the early years). These individuals held in common their belief in the value in getting to the truth of stories widely circulated online and the pleasure in the chase for this truth. The methods that were employed were nothing more than a healthy dose of skepticism, clear thinking, and basic library research – and occasionally talking to experts.

   In particular, when AFU members were studying urban legends, they disavowed being folklorists; their research into the truth behind particular urban legends was not colored by the methods and theories of academic folklore studies. In fact, the professional folklorists did not dominate participation on the AFU website. As we shall see in Chap. 4, the folklorists interested in urban legends, or *contemporary legends* as they called them, organized themselves instead around the International Society for Contemporary Legend Research, its newsletter (*FOAFTale News*), and its scholarly journal (*Contemporary Legend*).

   Similarly, when AFU became engaged in rumors surrounding major public events such as the Bush-Gore election in 2000 and the 9/11 terrorist attacks in 2001, it did not rely on the theories or methods of political scientists or other social scientists; instead, the individual members continued to base their findings in native rational thought, library research, and some legwork to track down the origins of stories. Smart and persuasive speech mattered; powerful academic research methods did not.

---

[34] By her count, at the time of her separation from snopes, she had written 1905 articles for the website. Dean (2017) notes that David was increasingly changing the website, writing Barbara out of the history of snopes.

[35] Folkenflik (2017). The details of these legal battles are messy. For more information, see Bruno (2017); Funke (2018); Masnick (2017a, b); Van Grove (2017).

When David and Barbara Mikkelson broke off from AFU to form snopes, over the long run it was a case of turning a hobby into a business. (At first, it was a case of turning a hobby in which they actively participated but did not control into a hobby they did control, based in a new technology – using the World Wide Web instead of Usenet. But soon this personal hobby evolved into a business.) What did not change at snopes was the basic methodology used to determine the veracity of stories and rumors. These were the same methods that had been honed by the most active members of AFU.

However, there was one big change. In AFU, the standards applied to an individual member's analysis of a particular urban legend or political story was whether the member had taken a rational and critical stance, and done enough library research. The cost of failing to meet these standards was sharp criticism within the group; the group could be quite mean-spirited in dealing with newbies and others who were insensitive to the community's culture – even trolling to get these people to behave outside the norms of the organization so that there was a reason to ostracize them. AFU was all about using strong social norms to preserve the hobby culture of AFU.

snopes was subject to three new standards of performance. First, national politics was big and serious business; and many partisan outsiders were much more interested in being persuasive than in uncovering the truth. When snopes contradicted the message of these partisans, it came under attack, as did other fact-checking sites – especially from conservative political interests. These attacks (e.g. "snopes has a liberal bias") were often fierce, often difficult to counter both because of their vagueness and because they relied on emotion rather than rational argumentation. In response, snopes tried to be explicit and transparent about its practices. Its website detailed every stage of the process in selecting, analyzing, and writing up its findings about a particular urban legend or political claim.

However, given that snopes had laid bare its process – and that it was one of native reasoning and ethical journalistic practice rather than driven by the standards of any academic discipline – this opened up snopes to criticisms from academics. These criticisms, such as those from information scholar Kalev Leetaru, asked both about whether snopes actually follows its own stated policies and also whether this kind of native rational approach is sufficient, or whether more powerful theories and methods from the academic disciplines need to come into play.

The fact that snopes runs as a business introduced a third set of standards, which did not apply to AFU, which was run as a hobby. Now, snopes had to worry about the independence of its opinions from its financial backers, meeting payrolls for its employees, the massive costs of advertising and website maintenance, and control over the company from partners. Some of these business risks have become clear since Barbara Mikkelson sold her shares to outsiders and David has been involved in lengthy and expensive court battles over the business aspects of snopes.

It is apparent that David Mikkelson's identity is closely wrapped up with snopes – his *nom de plume* while participating on AFU and the name of his company. The court costs over ownership of snopes may in the end cost him more than $2 million, but he seems determined to pay whatever it takes to retain control of this company. Given the immediate success of his first gofundme campaign, he may well be able to raise the funds – if he does not already have them – to continue the fight for his company in the courts. But this legal battle also makes one realize how fragile the continuation of snopes is – resting on a single person and a dozen employees rather than being backed by an organization with a long history and deep pockets.

Finally, we might ask, how do the roles of AFU and snopes fit into the broader scope of America's tug of war between truth and falsehoods, between overt manipulation of facts in support of a point of view or organization and the desire of the public for truthful certainty? In an earlier study, we documented cases of truth in battle with misinformation and observed three behaviors in the United States (Cortada and Aspray 2019). First, urban legends, lies, rumors, and misinformation of the types these two websites monitored had existed since the early nineteenth century. Second, truth seekers used the information technologies of their day to distinguish falsehoods from actual facts. Third, these efforts were highly fragmented, thus accessible only to pockets of citizens. The two case studies in this chapter demonstrate that our three findings about earlier behaviors, and the values that motivated them, continued after the arrival of the Internet.

There also were some differences, however. The most obvious was that participants employed humor and game-like behavior while debunking, yet at the same time were as serious about their work as one might find among academic researchers. Second, by using the Internet as their vehicle for deliberating and delivering information, they were able to leverage a larger pool of researchers and experts and to reach much larger audiences than others could in any earlier period in American history. Those two features of their work increased their visibility, thus their potential influence on users of the Internet. The work of these two organizations also reflected a growing interest – indeed need – in American society for validating truth. By the end of the 2016 presidential campaign, news organizations had demonstrated to the public that they too were doing similar things as AFU and snopes, but within the context of their more traditional roles in journalism. Such news outlets as the *LA Times, New York Times,* and *Washington Post* set up teams of reporters to identify lies, largely by politicians, less about other matters, publishing their findings both in print and online editions of their publications. Thus, such outlets in combination with a growing list of Internet-based truth seekers were creating a complex web of fact overseers in American society far more intricate and sophisticated than has existed in earlier times.

While some of the fact checking by these two Internet-based organizations may appear even trivial, the concerns over urban legends and truth could be very serious. No modern case so proves this point than what happened with respect to the events of September 11, 2001. For that reason, we devote the next chapter to this topic. It is a serious and complex story, made more urgent by the fact that one could reasonably expect future national crises to be muddled by fake news and settled facts, and by various constituencies packaging information in support of their parochial interests. We now turn to 9/11 to broaden our understanding of modern fact-checking activities.

References

Adams, Cecil. 1984. *The Straight Dope*. Chicago: Chicago Review Press.
———. 1988. *More of the Straight Dope*. New York: Ballantine Books.
———. 1994. *Return of the Straight Dope*. New York: Ballantine Books.
Baker, Paul. 2001. Moral Panic and Alternative Identity Construction in Usenet. *Journal of Computer-Mediated Communication* 7. https://academic.oup.com/jcmc/article/7/1/JCMC711/4584243. Accessed 18 July 2018.
Baym, N. 1995. From Practice to Culture on Usenet. *The Sociological Review* 42 (S1): 29–52.
———. 1999. *Tune In, Log On: Soaps, Fandom, and Online Community*. Thousand Oaks: Sage.
Blank, Trevor J., and Robert Glenn Howard. 2013. *Tradition in the Twenty-First Century: Locating the Role of the Past in the Present*. Boulder: University Press of Colorado.
Bond, Paul. (2002). Web Site Separates Fact from Urban Legend. *SFGate,* September 7. http://www.sfgate.com/entertainment/article/Web-site-separates-fact-from-urban-legend-2800717.php. Accessed 18 May 2017.
Bronner, Simon J. 1990. *Piled Higher and Deeper*. Atlanta: August House.
Bruckman, Amy. 1994. Truth or Legend? *Wired,* February 1. https://www.wired.com/1994/02/electric-word-11/. Accessed 2 July 2018.
Bruno, Bianca. 2017. Fact-Checker Snopes' Owners Accused of Corporate Subterfuge. *Courthouse News Service.* https://www.courthousenews.com/fact-checker-snopes-owners-accused-corporate-subterfuge/. Access 25 Sept 2018.
Brunvand, Jan Harold. 1981. *The Vanishing Hitchhiker: American Urban Legends and Their Meaning*. New York: W.W. Norton.
———. 1984. *The Choking Doberman and Other 'New' Urban Legends*. New York: W.W. Norton.
———. 1986. *The Mexican Pet: More New Urban Legends and Some Old Favorites*. New York: W.W. Norton.
———. 1993. *The Baby Train and Other Lusty Urban Legends*. New York: W.W. Norton.
———. 1999. *Too Good to be True: The Colossal Book of Urban Legends*. New York: W.W. Norton.
———. 2001. Folklore in the News (And, Incidentally, on the Net). *Western Folklore* 60 (1): 47–66.
———. 2002. *Encyclopedia of Urban Legends*. New York: W.W. Norton.
———. 2004. *Be Afraid, Be Very Afraid: The Book of Scary Urban Legends*. New York: W.W. Norton.
Burger, Peter. 2002. The Invisible Field-worker: Contemporary Legend Research on Usenet. *FOAFTale News,* No. 53, December 2002. http://www.folklore.ee/FOAFtale/ftn53.htm#abstracts. Accessed 22 June 2018.
Burnett, Gary, and Laura Bonnici. 2003. Beyond the FAQ: Explicit and Implicit Norms in Usenet Newsgrooups. *Library and Information Science Research 25*: 333–351.
Church, Bob. 2018. Oral history interview conducted by Alexis de Coning for this book.
Cortada, James, and William Aspray. 2019. *Before Fake Facts: The Long History of Lies and Misrepresentations in American Public Life*. Lanham: Rowman & Littlefield.

Dean, Michelle. 2017. Snopes and the Search for Facts in a Post-Fact World. *Wired,* September 20, https://www.wired.com/story/snopes-and-the-search-for-facts-in-a-post-fact-world/. Accessed 25 Sept 2018.

Dégh, Linda. 1994. *American Folklore and the Mass Media.* Bloomington: Indiana University Press.

DeGroot, Jocelyn M. 2011. Truth in Urban Legends? Using Snopes.com to Teach Source Evaluation. *Communication Teacher* 25 (2): 86–89.

Donovan, Pamela. 2002. Crime Legends in a New Medium: Fact, Fiction and Loss of Authority. *Theoretical Criminology* 6 (2): 189–215.

———. 2004. *No Way of Knowing: Crime Urban Legends, and the Internet.* New York: Routledge.

Dundas, Alan. 1987. *Cracking Jokes: Studies of Sick Humor Cycles and Stereotypes.* Berkeley: Ten Speed Press.

Dundas, Alan, and Carl Pagter. 1975. *Urban Folklore from the Paperwork Empire.* Bloomington: American Folklore Society.

Emerson, S. L. 1983. Usenet: A Bulletin Board for Unix Users. *Byte* 8(10, October): 219–236.

Fader, Carole. 2012. Fact Check: So Who's Checking the Fact Finders? We Are. *Forida Times-Union,* September 28. http://jacksonville.com/news/metro/2012-09-28/story/fact-check-so-whos-checking-fact-finders-we-are. Accessed 18 May 2017.

Farish, Terry. 1994, October. Choice Interviews: Linda Degh. *Choice* 32(2): 249–251.

Faulkner, William. 1940. *The Hamlet.* New York: Random House.

———. 1957. *The Town.* New York: Random House.

———. 1959. *The Mansion.* New York: Random House.

Folkenflik, David. 2017. Who's the True Boss of Snopes? Legal Fight Puts Fact-Check Site at Risk. *National Public Radio,* July 26. https://www.npr.org/2017/07/26/539576135/fact-checking-website-snopes-is-fighting-to-stay-alive. Accessed 2 July 2018.

Frentzen, J. 1997. Usenet's Underbelly: Where the Wild Users Are. *PC Week* 14(48, November 17): 35.

Funke, Daniel. 2018. Snopes Has Its Site Back. But the Legal Battle Over Its Ownership Will Drag on for Months. *Poynter,* March 20. https://www.poynter.org/news/snopes-has-its-site-back-legal-battle-over-its-ownership-will-drag-months. Accessed 25 Sept 2018.

Furr, Joel. 1995. The Ups and Downs of Usenet. *Internet World* 6(11): 58–61.

Goldstuck, Arthur. 1994. *The Ink in the Porridge: Urban Legends of South African Elections.* London: Penguin.

Goodman, Alana. 2016. *EXCLUSIVE: Facebook 'Fact Checker' Who Will Arbitrate on Fake News Is Accused of Defrauding Website to Pay for Prostitutes – And Its Staff Includes an Escort-Porn Star and 'Vice Vixen Domme',* 21 December. http://www.dailymail.co.uk/news/article-4042194/Facebook-fact-checker-arbitrate-fake-news-accused-defrauding-website-pay-prostitutes-staff-includes-escort-porn-star-Vice-Vixen-domme.html. Accessed 26 May 2017.

goodnature. 2016. Snopes (Snopes.com). *Truth Wiki.* http://www.truthwiki.org/snopes-snopes-com/. Accessed 26 May 2017.

Hardy, H. E. 1993. *The History of the Net.* Master's thesis, School of Communications, Grand Valley State University, Allendale, MI.

Hauben, Michael. 1997. Culture and Communication – The Interplay in the New Public Commons: Usenet and Community Networks. In *An Ethical Global Information Society,* ed. J. Berleur and D. Whitehouse, 197–202. London: Chapman and Hall.

Higgins, Andrew, Mike McIntire, and Gabriel J. X. Dane. 2016. Inside a Fake News Sausage Factory: 'This Is All About Income. *New York Times,* November 25. https://www.nytimes.com/2016/11/25/world/europe/fake-news-donald-trump-hillary-clinton-georgia.html?action=click&contentCollection=Technology&module=RelatedCoverage&region=EndOfArticle&pgtype=article. Accessed 18 May 2017.

Hochman, David. 2009. Rumor Detectives: True Story or Online Hoax? *Reader's Digest,* March 10. http://web.archive.org/web/20090310014243/http://www.rd.com:80/your-america-inspiring-people-and-stories/rumor-detectives-true-story-or-online-hoax/article122216.html. Accessed 18 May 2017.

Isaac, Mike. 2016. Facebook Considering Ways to Combat Fake News, Mark Zuckerberg Says. *New York Times,* November 19. https://www.nytimes.com/2016/11/20/business/media/facebook-considering-ways-to-combat-fake-news-mark-zuckerberg-says.html. Accessed 18 May 2017.

Klintberg, Beng af. 1990. *Rattan i Pizzan* [The Rat in the Pizza]. Basingstoke: Pan

Kohn, Alfred. 1990. *You Know What They Say…The Truth About Popular Beliefs*. New York: Harper.

Krassner, Paul, ed. 1984. *Best of the Realist*. Philadelphia: Running Press.

Lee, H. 2002. "No Artificial Death, Only Natural Death": The Dynamics of Centralization and Decentralization of Usenet Newsgroups. *The Information Society* 18 (5): 361–370.

Leetaru, Kalev. 2016. The Daily Mail Snopes Story and Fact Checking the Fact Checkers. *Forbes,* December 22. https://www.forbes.com/sites/kalevleetaru/2016/12/22/the-daily-mail-snopes-story-and-fact-checking-the-fact-checkers/#1b36d9ea227f. Accessed 26 May 2017.

Masnick, Mike. 2017a. Fact Checking Snopes on Its Own Claims of Being 'Held Hostage' by a 'Vendor': Well, It's Complicated. *Techdirt,* August 1. https://www.techdirt.com/articles/20170731/11351837890/fact-checking-snopes-own-claims-being-held-hostage-vendor-well-complicated.shtml

———. 2017b. The Snopes Fight Is Even Way More Complicated than We Originally Explained. *Techdirt,* August 15. https://www.techdirt.com/blog/?company=proper+media. Accessed 25 Sept 2018.

Nissen, Beth. 2001. Hear the Rumor? Nostradamus and Other Tall Tales. *CNN.com.* http://www.cnn.com/2001/US/10/03/rec.false.rumors/index.html. Accessed 18 May 2017.

North, Tim. 1994. *The Internet and Usenet: Global Computer Networks: An investigation of Their Culture and Its Effects on New Users*. Unpublished Master's Thesis, Curtin University of Technology, Perth, Australia. https://web.archive.org/web/20001206042500/http:/www.vianet.net.au/~timn/thesis/index.html. Accessed 3 Aug 2018.

Novak, Viveca. 2009. Snopes.com. *Ask FactCheck.Org,* April 10. http://www.factcheck.org/2009/04/snopescom/. Accessed 18 May 2017.

Opie, Iona, and Peter Opie. 1972. *The Lore and Language of School Children*. Oxford: Clarendon Press.

Pogue, David. 2010. At Snopes.com, Rumors Are Held Up to the Light. *New York Times,* July 15. http://www.nytimes.com/2010/07/15/technology/personaltech/15pogue-email.html. Accessed 17 May 2017.

Poundstone, William. 1983. *Big Secrets*. Fort Mill: Quill.

Richer, Duncan. 1998. *The Den of Iniquity – Alt.Folklore.Urban,* April 4. http://www.chiark.greenend.org.uk/~dricher/afu.html. Accessed 17 May 2017.

Seipp, Catherine. 2004. Where Urban Legends Fall. *National Review,* July 21. https://web.archive.org/web/20040812075515/http://www.nationalreview.com/seipp/seipp200407210830.asp. Accessed 18 May 2017.

Sheidlower, Jesse, ed. 1995. *The F-Word*. New York: Random House.

Smith, Paul. 1984. *The Complete Book of Office Mis-Practice*. London/Boston: Routledge/Kegan Paul.

Smith, Paul, and Gillian Bennett. 1987. *Perspectives on Contemporary Legend*. Sheffield: Sheffield Academic Press.

———. 1988. *Monsters with Iron Teeth*. Sheffield: Sheffield Academic Press.

snopes.com. 2017a. *About Snopes.com.* http://www.snopes.com/about-snopes/. Accessed 19 May 2017.

———. 2017b. *Snopes.com Staff.* http://www.snopes.com/snopes-staff/. Accessed 19 May 2017.

———. 2017c. *Transparency: Corrections Policy.* http://www.snopes.com/corrections-policy/. Accessed 22 May 2017.

———. 2017d. *Transparency: Methodology*. http://www.snopes.com/methodology/. Accessed 19 May 2017.

———. 2017e. *Transparency: Ownership and Revenue*. http://www.snopes.com/ownership-and-revenue/. Accessed 22 May 2017.

———. 2017f. *Transparency: Ratings System*. http://www.snopes.com/ratings/. Accessed 22 May 2017.

———. 2017g. *Transparency: Sources*. http://www.snopes.com/sources/. Accessed 22 May 2017.

———. 2017h. *Transparency: Topic Selection*. http://www.snopes.com/topic-selection/. Accessed 19 May 2017.

———. 2017i. *Glossary*. http://www.snopes.com/glossary/. Accessed 22 May 2017.

Stelter, Brian. 2010. Debunkers of Fictions Sift the Net. *New York Times,* April 4. http://www.nytimes.com/2010/04/05/technology/05snopes.html. Accessed 18 May 2017.

Streitfeld, David. 2016. For Fact-Checking Website Snopes, a Bigger Role Brings More Attacks. *New York Times,* December 25. https://www.nytimes.com/2016/12/25/technology/for-fact-checking-website-snopes-a-bigger-role-brings-more-attacks.html?_r=0. Accessed 18 May 2017.

Tepper, M. 1997. Usenet Communities and the Cultural Politics of Information. In *Internet Culture*, ed. D. Porter, 9–54. New York: Routledge.

Tindall, Bruce, and Mark Watson. 1991. *Did Mohawks Wear Mohawks? And Other Wonders, Plunders, and Blunders*. New York: William and Morrow.

———. 1994. *How Does Olive Oil Lose Its Virginity?* New York: William and Morrow.

Tomblin, Paul. 2018. Oral history interview conducted by Alexis de Coning for this book.

Turner, Patricia. 1993. *I Heard It Through the Grapevine: Rumor in African-American Culture*. Berkeley: University of California Press.

Turner, Tammara Combs, et al. 2005. Picturing Usenet: Mapping Computer-Mediated Collective Action. *Journal of Computer-Mediated Communication* 10(4). https://academic.oup.com/jcmc/article/10/4/JCMC1048/4614533. Accessed 18 July 2018.

Van der Linden, Peter. 1989. *The Official Handbook of Practical Jokes*. New York: Signet.

———. 1991. *The Second Official Handbook of Practical Jokes*. New York: Signet.

Van Grove, Jennifer. 2017. Snopes.com Prevails in Tentative Court Ruling Over Finances, Ownership. *Phys.org,* August 4. https://phys.org/news/2017-08-snopescom-prevails-tentative-court-ownership.html. Accessed 25 Sept 2018.

Walker, Rob. 2016. How the Truth Set Snopes Free. *The Webby Awards,* October 19. http://www.webbyawards.com/lists/how-the-truth-set-snopes-free/. Accessed 17 May 2017.

Chapter 3
Urban Legends and Rumors Concerning the September 11 Attacks

On September 11, 2001 2996 people died in the United States in four coordinated suicide attacks by 19 members of the Islamic extremist group al-Qaeda and more than 6000 others were injured. The victims included people on four commercial airplanes hijacked by the terrorists, others in the World Trade Center in New York City and in the Pentagon Building in Arlington, Virginia when three of the planes crashed into them, people in the fourth plane which crashed into a field in Pennsylvania after the crew and passengers overtook the terrorists, and more than 400 policemen and firefighters who lost their lives in the rescue efforts. The attacks were masterminded and funded by a wealthy Saudi, Osama bin Laden, presumably in retaliation to American support of Israel and the American military presence in the Middle East.[1]

The terrorist attacks led to a widespread loss of a feeling of safety within the United States, together with a heightened sense of fear and increased prejudice against people of Arab origin. The attacks also led to a US military initiative, Operation Enduring Freedom, which began less than a month later, with the goal of destroying the Taliban, in particular bin Laden's terrorist network. While the power of the Taliban was greatly diminished by the end of 2001, vestiges of the terrorist network remained and bin Laden eluded Western troops for almost 10 years, until he was killed in a raid in Pakistan in 2011. The military operation was significantly reduced after his death.[2]

---

[1] On the writings of bin Laden, see Ibrahim (2007).

[2] For overall information on the terrorist attacks, see the official government report (Kean (2011)) as well as Bergen (2011); and Wright (2006); on the terror associated with the attacks, see Pyszczynski et al. (2003); on the relationship between political acts and the psychology of terror, see Huddy and Feldman (2011); and Morgan (2009); on memorialization and grieving, see Aronson (2016); Linenthal et al. (2013); Laqueur (2015); and Rosenblatt (2015); and on reactions to 9/11 as a cultural phenomenon, see Melnick (2009). Also see the Library of Congress September 11, 2001 Web Archive, which includes the snopes.com collection and The Avalon Project of the

© Springer Nature Switzerland AG 2019
W. Aspray, J. W. Cortada, *From Urban Legends to Political Fact-Checking*, History of Computing, https://doi.org/10.1007/978-3-030-22952-8_3

3.1   The Creation of Rumors and Legends in Response to 9/11

It would be difficult to overstate the anxiety felt by ordinary Americans as a consequence of the 9/11 attacks. Most Americans had believed they lived in a relatively safe place, but the demonstrated possibility of a terrorist attack where they lived or worked created tremendous consternation. No longer was living a good and just life assurance against violence or death to one's self or one's friends and family. As folklorist Linda Degh explained, urban legends are a coping mechanism for "the anxieties of ordinary people."[3] They provide a way for people to manage their fear by creating a narrative about a complex, dangerous, and incomprehensible situation, giving it a meaning that one can learn from, or at least cope with. This coping strategy, rightly or wrongly, can identify people unlike themselves – "others" – who can be blamed for this complex and dangerous situation.[4]

Numerous urban legends and rumors appeared in the days following 9/11, and continued to be created and circulated for the next two decades. The frequency diminished somewhat after the first year and again after the death of Osama bin Laden, but these stories never completely died out; some appear even today.

Later in this chapter, we analyze 176 legends and rumors related to 9/11 that were evaluated by snopes.com between 2001 and 2011 – most of them circulated online.[5] Some common themes run throughout these legends. Some involve *foreknowledge*, for example a "grateful terrorist" using his insider knowledge to warn an American of an impending terrorist attack in response to some kindness the American had shown to him. Many of the legends have a *conspiracy* theme, for example about all the Jews who avoided the World Trade Center on 9/11, or about the fact that the federal government knew about the coming attacks but did nothing to stop them for any of a variety of reasons. Some of the legends are instances of

---

Yale Law School. For personal accounts of those involved in the attacks, see DiMarco (2007); Dwyer and Flynn (2004); and Smith (2003). For fiction inspired by the attacks, see DeLillo (2008); Foer (2006); and Waldman (2011); and for an overview of literature since 9/11 see Gray (2011).

[3] Degh (2001), as quoted in Smith et al. (2010). Also see Gerould (1908/2000).

[4] Smith et al. (2010). Also see Lindahl (2009) on the healing power of folklore images.

[5] One might question how online behavior and offline behavior in reaction to the 9/11 attacks compared. Dutta-Bergman (2006) argues that there was strong channel complementarity: "individuals who posted their thoughts in online communities were significantly more likely to attend a meeting to discuss the attacks, to volunteer in relief efforts, to write about their views to a newspaper or other news organization, and to sign a petition regarding the attacks as compared to those other individuals who did not post their thoughts in online communities. No significant differences were found in the realm of attending religious services or donating blood. This may be a reflection of the noncommunicative nature of these activities.... The common theme that joins online and offline community participation is the individual's orientation toward participating in the community with respect to an important crisis."

pareidol, the process of identifying some human trait in a complex pattern, such as seeing demonic faces in the smoke of the World Trade Center. A number of the legends express *religious* themes, such as an undamaged Bible found in the wreckage of one of the crashed planes. A few of these legends discuss *heroes*, such as actor Steve Buscemi, who worked alongside firefighters clearing the rubble at the World Trade Center; or *villains*, such as the Arab employees at a Dunkin Donuts restaurant who allegedly cheered when the planes hit the Twin Towers.[6]

9/11-type urban legends were not confined to the United States. For example, in Germany and the Netherlands a version of the Grateful Terrorist legend circulated widely before Christmas 2002. It took the form of an impending attack at the Christmas fairs that are traditionally held in many European town centers. One urban legend – about people who became ill from eating in a Middle Eastern restaurant because the garlic sauce had been contaminated with semen – which had circulated occasionally during the 1990s, was revived after the 9/11 terrorist attacks, and revived yet again after the March 2004 terrorist attacks in Madrid (Meder 2009). In 2003, the Smiley Gang rumor was particularly widespread across the Netherlands, involving approximately 100 online threads and 1000 people. The Smiley Gang was allegedly a Moroccan youth gang that would mutilate a victim's face to introduce a permanent smile if she – or occasionally, he – would not submit to be raped. Variations of the Smiley Gang legends identified the victim as an adolescent girl, a couple, or younger children; and sometimes included a coating of HIV on the knife blade. This rumor appeared in the year following the murder of anti-Islamic politician Pim Fortuyn during the campaign leading to the Dutch national elections, resulting in heightened anti-Islamic tensions in the country (Burger 2009).

9/11-style urban legends appeared for the first time long before the 9/11 attacks. The Grateful Terrorist legend is more than 2000 years old, e.g. appearing in the Hebrew tale of Tobit in 721 BC and then later retold in Cicero's tale of Simonides in *De Divinatione* in 44 BC (Smith et al. 2010). Legends similar to the Smiley Gang legend had appeared in 1915, during the First World War, at that time directed at the Germans as villains, and have reappeared many times since 1950 in England, Scotland, France, Belgium, and the United States (Meder 2009; Burger 2009).

During the Second World War, the US government created rumor management clinics to debunk the numerous legends that had begun to spread.[7] Realizing that the Pearl Harbor attacks of December 7, 1941 had generated many rumors and legends, the Texas folklorist and ethnomusicologist Alan Lomax organized a project for people to collect stories about Pearl Harbor and the US declaration of war (Goldstein 2009). Stories from this time included a number of foreknowledge legends.[8]

---

[6] This categorical scheme of 9/11 legends emerged from our analysis of the snopes.com 9/11 archive, but also see Heimbaugh (2001).

[7] Goldstein (2009). Also see Behe (1988).

[8] Bonaparte (1947), as quoted in Goldstein (2009).

3.2   snopes and 9/11

Almost immediately after the 9/11 terrorist attacks, rumors began to circulate. They persisted not merely for a few weeks or months, but for years. snopes tracked and studied many of these 9/11-related rumors and (at the time of this writing) maintains an archive of 176 rumors collected between 2001 and 2011. snopes clearly had the pulse of these rumors in hand. For example, it received more than 700 inquiries within a 12-h period regarding one rumor about a boyfriend who advised his girlfriend not to travel on September 11, 2001 and not to visit shopping malls on October 31, 2001, the day of Halloween (Goldstein 2009). Our summary of the 9/11 rumors collected by snopes can be found at https://scholar.colorado.edu/infosci_facpapers/4/ . This section provides our own analysis of the 9/11 rumors and legends, based on the material appearing in snopes.

Table 3.1 provides a categorization of the 9/11-related rumors collected by snopes (for those rumors we were able to categorize clearly).

It is not surprising that the largest number of rumors concerned new acts of terrorism or suspicious behaviors. The 9/11 attacks were totally unexpected, well

Table 3.1 9/11-Related legends and rumors collected by snopes, categorized by type

Category of rumor	Number
Follow-up Terrorist Attacks and Suspicious Behavior	25
Editorials, Articles, and Speeches	18
Curiosities and Human-Interest Stories	14
Memorials, TV Testimonials, and Celebrations	12
Stories about Osama bin Laden and Afghanistan	12
Inappropriate and Unpatriotic Behavior	11
Business Practices as Related to 9/11 and its Aftermath	9
Stories of Foreknowledge	8
Pareidol, Folk Explanations, and Numerology	8
Religion and Miracles	8
Heroes and Exemplary Behavior	6
Celebrities and 9/11	6
Boneheaded Government Actions	5
Historical Precedents and Prophecies	4
"Celebrating Arabs" Stories and Reprisals	4
Practices of Terrorists	4
Preventive Practices against Terrorism	3
Visual Images	3
Conspiracy Theories	1

planned, and well financed, and the leader of the terrorist group Osama bin Laden remained at large for a decade despite considerable efforts to capture or kill him. If such an elaborate plan could be carried out on September 11, there was no telling when and where the terrorists might strike again. Especially in 2001 and 2002, but continuing over the next decade, rumors appeared concerning prospective attacks to bridges in California and shopping malls on New York's Long Island, suitcase bombs in the New York City subway, and a nuclear bomb smuggled into the United State by al-Qaeda. Some of these rumors took the form of the Grateful Terrorist legend.

Other rumors in this category included incidents or behaviors that, under other circumstances, might not have elicited concern, but given the fear generated by the 9/11 attacks, made people worry. Examples included missing Ryder rental trucks, unexplained crop duster sprayings, a cyanide truck hijacking in Mexico, five men illegally crossing the US border from Canada, and women in burkhas buying cell phones and texting in a cinema.

Many people across the United States and around the world wrote editorials or articles, or gave speeches, about terrorism in the months after 9/11. There was a visceral hunger for information and wisdom about these terrorists who wanted to destroy the American way of life, and a desire for reassurance that the average American citizen was safe or at least could figure out how to protect him- or her- self. When someone offered wisdom or opinion, there was a natural desire to share these offerings online. However, communication channels were often imperfect: metadata about the provenance of a talk was often lost, and a message might become garbled. Thus, it was common for incomplete and incorrect information about these editorials, articles, and lectures to be transmitted through email and on websites. snopes investigated many of these cases, which assumed a wide variety of forms and emanated from a wide variety of individuals. They included a patriotic speech by Senator John McCain, an interview with a filmmaker who had studied terrorism in the Middle East discussing the psychology of terror, a report from musician Charlie Daniels about what he saw when he visited the American prison in Guantanamo Bay, Cuba, an essay by spy thriller author John Le Carre, an army veteran describing the nature of chemical and nuclear warfare, a first-person account of the escape from a high floor of the World Trade Center, an anti-Bill Clinton diatribe from a firefighter, a letter to terrorists from an "ordinary" grandfather, and an editorial written from the perspective of an Arab-American citizen.

The 9/11 terrorist attacks affected people in many different ways – all of them human and personal. This led to many human-interest stories being passed around through email and blogs – some of them true, but many of them satisfying some human need rather than being based in fact. These stories took a variety of forms: two were about people who survived the 9/11 attacks only to die within a few months – one in a ferry crash, the other in a plane crash. Some were heartwarming stories, for example about the mother cat and her kittens that survived the World Trade Center crash and were living in the rubble, or the story of the women who named her newly born son after the firefighter who rescued her. Others were gruesome, such as the story of finding airline passengers still strapped in their seats in an apartment building near the World Trade Center; or sad and incredulous, such as the story of the son of one of the World Trade Center survivors joining the Taliban a few months later. Some were appreciated for their colorful language, such as The Binch (a parody of Dr. Seuss's Grinch story focused on Osama bin Laden) or President George W. Bush's remark that he would not fire a two-million-dollar missile at a $10 tent only to hit a camel in the butt. Others included a supposed connection between baseball star Jackie Robinson and one of the passengers who fought the terrorists on the plane that crashed in Pennsylvania, the Canadian policeman who was suspended from his job for traveling to New York City to help with the rescue effort, and the fact that hospitals around New York staffed up 9 months after 9/11 in expectation of a large run on their maternity facilities. A personal favorite was the Philadelphia company that placed an ad that said, "we would prefer to serve one terrorist to serving 1000 Jews", which seemed appalling until one realized that the organization was a Jewish mortuary and the quip was intended to be humorous but patriotic.

There were numerous efforts across the United States – and especially so in New York City – to demonstrate patriotism and solidarity through memorials, celebrations, television testimonials, and institutional and individual websites for the brave men and women who had worked in the 9/11 rescue or who were in the armed services protecting the country from terrorists.[9] These expressions took many different forms: a metal sculpture donated by Russia and a Canadian artist's ice sculpture representing the World Trade Center firefighters; a petition to the United Nations, a Patriot Day proclamation, and a federal declaration for a National Day of Reconciliation; a candlelight vigil and an attempt to get people across the nation to turn on their automobile headlights at a certain time so that they could be seen from space; and a Budweiser ad titled "Respect" shown during the 2002 Super Bowl.[10] People heard rumors of these various efforts and wanted to know about them. As above, sometimes the message became garbled. While these rumors overwhelmingly had a pro-US celebratory orientation, there was one rumor about a pro-terrorist mural hanging publicly in Iraq.

The American public showed considerable interest in Osama bin Laden, as the mastermind of the 9/11 attacks and as someone able to elude American forces for years. Many rumors circulated about him and what was going on in Afghanistan. These stories took many forms: he was dying of kidney failure; he had been sighted in Utah; his family was able to escape the United States shortly after 9/11 on private planes with the cognizance of the U.S. government; he owned various businesses, including the Snapple distributorships in the Middle East that should be boycotted; and even a humorous memo that some people took as fact in which he chastised his colleagues for stealing his Cheez-Its.

---

[9] On the healing power of art and other creative responses to the 9/11 attacks, see Zeitlin and Harlow (2001). On how memorializing online differed from memorializing in person, see Foot et al. (2005). On memorialization at the Pentagon, see Greenspan (2003) and Yocom (2006).

[10] See Jackson (2005) for an interesting account of how the Hallmark company remained faithful to its corporate values in giving psychological aid after the 9/11 attacks.

Patriotism was running high after 9/11, and people were particularly sensitive to the practices of companies and individuals that they felt might not be entirely patriotic in spirit. Outrage was expressed over a number of incidents, and email flew around the country reporting these behaviors: a Starbucks near the World Trade Center charging exorbitant prices for bottled water to firefighters; September 11 charitable funds allegedly being used to defend terrorism suspects in court; removal of God Bless America banners in public schools in several cities; removal of American flags from fire trucks in Berkeley, California; Charlie Daniels not being permitted by concert promoters to sing certain patriotic songs in a concert; and *Time* magazine considering bin Laden for its man-of-the-year award.[11] Perhaps the tackiest story that circulated concerned a television ad for a San Antonio, Texas mattress firm, which alluded to 9/11 themes in its hard-pitch, late-night-style commercial, ending with two towers of mattresses crashing down at the end of the commercial.

Various companies decided they needed to adjust their business practices because of the sensitivities created by the 9/11 attacks. Starbucks removed a poster for a new summer cold drink, because the dragonfly in the poster appeared to be crashing into the two coffee cups. Pepsi eliminated one of its soft drink can designs for similar reasons. The national radio broadcast company, Clear Channel, identified a set of popular songs that it discouraged its affiliates from playing because they could evoke memories of the terrorist acts or incite violence. Amazon ended a partnership with the pro-Palestinian company, Intifada. The Red Cross received widespread criticism both for taking high overhead from post-9/11 donations and for spending these funds on a variety of projects that extended beyond 9/11-related projects. Spirit Airlines, however, decided to take advantage of the 9/11 event by giving free flights on the anniversary of the terrorist attacks. Various automobile manufacturers decided to make substantial donations to 9/11 funds.

Many rumors circulated about groups of people having foreknowledge of the 9/11 events. There were invariably reasonable alternative explanations to these claims of foreknowledge, although one such rumor about 4000 Israelis staying away from the World Trade Center on 9/11 was so persistent that the US State Department eventually issued a public denial. These foreknowledge rumors took various forms: a prediction on 9/10 by a Texas schoolchild that World War III would begin the following day; a Dallas religious leader telling his congregation to stock up on supplies shortly before 9/11; registration of domain names such as horrorinnewyork.com in advance and selling them off after the attacks; and the absence of taxis at the World Trade Center the morning of the attacks.

The Internet was alive with pseudo-scientific reports and explanations connected to the terrorist attacks. Some were instances of *pareidol*, for example, finding images of Satan in the smoke from the World Trade Center, or in finding that if one typed the letters NYC in Microsoft's Wingdings font in its Windows 3.1 release, the output was a skull-and-crossbones, Jewish star, and thumbs up glyph. There were

---

[11] *Time* had previously made this award to Hitler and Stalin, so it was clearly about people who changed the world but not necessarily for the better; however, under intense pressure, *Time* deciding not to give the award to Osama, in the end calling him a run-of-the-mill terrorist.

also examples of numerology, such as people observing the winning number combinations in the 9/11/2002 New York lottery or, later, in discovering that it was 911 days from the 9/11 attacks in New York and Washington to the 2004 terrorist attacks in Madrid, Spain. Others gave folk explanations related to the 9/1 bombings, e.g. some believed that a taunt of Osama bin Laden by a New York firefighter was the cause of a plane crash in the New York area several months after 9/11.

Not surprisingly, religion was a common theme in the 9/11 rumors that spread on the Internet, involving both religion as solace in complex times of trouble and as the retribution of a vengeful God. Some of these rumors involved signs of Divine presence, such as World Trade Center girders found in the shape of a cross in the rubble, or an unburned Bible in the Pentagon wreckage. Others called for the use of prayer, such as a rumor about the establishment of a National Prayer Team and a grass-roots effort to encourage people to pray for the soldiers sent to Afghanistan to capture bin Laden. There were also stories circulating about famous religious figures, such as Billy Graham's daughter chastising the American public for believing in God only during times of crisis, and the alleged speech by the Dalai Lama rousing people to become spiritual activists. There were also reports of conservative televangelists blaming the terrorist attacks on American liberalism.

A number of stories appeared lauding heroic actions related to 9/11. Examples included Outback Steakhouse donating food and serving steak dinners to US troops in Afghanistan, and the story of the people of Gander, Newfoundland, who took into their homes for a week the passengers on numerous flights from Europe grounded by the 9/11 attacks. Both of these stories were true; whereas there was only limited truth to the story that circulated concerning President George W. Bush being considered for the Nobel Peace Prize on account of his 9/11 response.

Another common theme – somewhat of the human-interest variety – were stories about connections of celebrities to the 9/11 events. These included the true story of Steve Buscemi (who had been a firefighter before becoming a movie star) joining in the World Trade Center disaster relief; the also true story of actor James Woods, who saw suspicious men perhaps practicing for the 9/11 attacks; rumors of a new Rambo film with Sylvester Stallone fighting the Taliban (not quite true); a Jackie Chan movie that was supposed to be filmed on the World Trade Center on September 11 (a greatly exaggerated story but with an element of truth); and an also-true rare public statement from Muhammad Ali about Islam as a religion of peace.

Boneheaded government action was another common theme in the 9/11 rumors. These were not the conspiracy theories in which public officials knew in advance that the terrorist activities were going to occur and let them happen anyway (discussed below), but instead were focused on the ineptitude of the government, especially the US federal government. These rumors took various forms: that the officials had captured and released Mohamed Atta, one of the 9/11 terrorist leaders (false); that the National Park Service restricted a patriotic rally following 9/11 (true, but for a good reason); that President Bush called on women to appear naked in public as a defensive measure, because of the religious sensitivities of the terrorists (ludicrous); that the Clinton Administration had been ineffective against terrorism; and that the various federal administrations had tied the hands of the military to fight terrorism.

In trying to gain some understanding of why the terrorist attacks occurred and some inkling as to whether they were really as unexpected as they seemed to be, Americans looked for understanding from the past in the form of both historical precedents and prophecies. One rumor (false) discussed the efforts of General John J. Pershing during the First World War in the Philippines to curb Muslim terrorism by killing Muslims and burying them in a place made unholy by the presence of pigs. Another false rumor was that White House aide Lt. Colonel Oliver North had warned Congress of the dangers of bin Laden during the Iran Contra hearings in 1987. Yet another false story involved text supposedly written by radio journalist Edward R. Murrow urging restraint in connection with the Japanese bombing of Pearl Harbor. The most widely circulated example of a purported prophecy concerned the sixteenth century French astrologer Nostradamus, whose ambiguous writings were interpreted in a way that "demonstrated" that he predicted the World Trade Center bombings.

One of the most common memes discussed in the scholarly literature about urban legends related to the 9/11 attacks concerns the "Celebrating Arabs" rumor. These stories are about people witnessing Arabs in the United States celebrating the terrorist attacks – at a Dunkin Donuts restaurant, or a 7-11 store, or at the Detroit area restaurant (The Sheikh, discussed later in this chapter) and the reprisals that were taken by ordinary Americans, such as the Budweiser delivery man removing Budweiser products from the 7-11 store or calls to boycott The Sheikh restaurant. Despite considerable scholarly discussion of this topic, snopes reported only two instances of this kind of rumor. Less discussed in the scholarly literature (but covered by snopes) are related rumors of Arab celebration outside the United States, such as the printing of T-shirts with 9/11 themes or dancing in the streets in Palestine after the planes crashed into the Twin Towers.

Two of the less common categories of rumors involved recognition of the practices of terrorists and dissemination of practices that could be adopted by American citizens to thwart terrorism. snopes reports on the following alleged terrorist practices that appeared in rumors: riding in the jump seat on a plane as practice for hijacking planes; using text messaging for coordination among hijackers; funding terrorist efforts through telemarketing scams; and collecting personal information about military personnel through bogus military-support websites. snopes reported on several rumors related to protecting against terrorist action: a talk given by a pilot on what passengers should do if their plane is hijacked, how to destroy anthrax with common household chemicals, and the possibilities and impacts of protecting against chemical attacks with the liberal use of duct tape.

Some rumors concerned visual images and whether they were original or had been digitally altered ("photoshopped" is the common term used here). The most important example is Tourist Guy, the image of a tourist on the observation deck of the World Trade Center whose back is turned away from an airplane that is about to crash into the building. (This image is discussed in detail later in this chapter.) snopes also reported on rumors concerning two other alluring images of the World Trade Center, both of which were innocently taken by amateur photographers. In both cases, they were pre-9/11 photos that had not been digitally manipulated and had little relevance to the 9/11 story.

The final class of 9/11 rumors discussed by snopes involves conspiracy theories, and this class includes only a single rumor. This rumor argues that the damage to the Pentagon was not the result of terrorists flying a plane into the building but instead was staged by the US government. While this rumor is false, what is perhaps most interesting about this category is that the scholarly community has devoted considerable attention to conspiracy theories in its discussion of 9/11 rumors; and based on the single instance of a conspiracy theory found by snopes, perhaps this scholarly attention to conspiracy theories is undeserved. We know, however, that conspiracy theories were rampant in other cases such as the assassinations of Abraham Lincoln and John Kennedy.[12]

3.3   Conspiracy Theories and the 9/11 Truth Movement

This final class of rumors provides a good transition to one of the most widely discussed topics of 9/11 rumor scholarship – conspiracy theories.[13] Conspiracy theories are "allegations that powerful people or organizations are plotting together in secret to achieve sinister ends through deception of the public" (Wood et al. 2013). Folklorist Veronique Campion-Vincent argues that while in earlier times evil was generally associated with religion (the evil other), in modern folklore there is often discussion of an evil elite, such as a government, and a complicit media beholden to these sources of elite power (Campion-Vincent 2003). She continues:

> Rather than explaining the attacks, such conspiracy theories partly deny their existence, and instead of raising methodological doubt adopt a generalized doubt close to a form of instant revisionism. Facts are not simply re-evaluated or contested, but denied practically at the same time as they happen and are told. Rather than focusing on the conspiracies of obscure outsiders, they mostly accuse the media, who are considered the heralds of the authorities. In this interpretative system, the media obstruct the truth, transmitting official versions of it whose undisclosed aim is actually to hide the conspiracy: as accomplices of the authorities, they are participating in the conspiracy… (Campion-Vincent 2003)

In one of the major scholarly papers on this topic, the sociologist Carl Stempel, the journalist Thomas Hargrove, and journalism professor Guido Stempel offer a psychological explanation for the frequency with which conspiracy theories appear:

> [The] literature suggests that conspiracy theories provide clarity of vision and clear targets for addressing the confusions, frustrations, and insecurities of living in contemporary societies which are characterized by rapid social change; a multiplicity of voices and interests; multi-level, multi-polar balances of power where those at higher levels maintain control through secrecy and controlling information; declining individual autonomy; increasing risk awareness associated with technological advances and "post-scarcity"

---

[12] See, for example, Cortada and Aspray (2019).

[13] On the history of conspiracy theories in American political thought, see a famous essay, Hofstadter (1964). For a contemporary account, see Hagen (2011).

conditions; high levels of social and geographic mobility; declining trust in national governments; and post-9/11 fears of terrorist/outsider threats.' (Stempel et al. 2007)

The irony here is that, at least proportionately, there has been so much scholarship about a topic that appeared at the bottom of the snopes list in terms of frequency of occurrence. In the case of 9/11 rumors, what is the nature of conspiracy theories? Some of these theories concern demographic classes of people who had some foreknowledge of the 9/11 attacks, such as the alleged mass avoidance by Jews of the World Trade Center on September 11; and this type of conspiracy theory often has racist overtones. The other main type of conspiracy theory involves the US government, possibly in collusion with state or local governments or foreign governments, who knew or suspected the attacks were coming but did nothing to stop them or alert the public. We consider in the remainder of this section three examples of conspiracy theory scholarship connected to the 9/11 terrorists. We follow those accounts with a discussion of one thread from alt.folklore.urban that questions the official account of the downing of Flight 93 in Pennsylvania before broadening into a discussion of the collapse of the Twin Towers.

Stempel, Hargrove, and Stempel conducted a national telephone survey of 1000 people in 2006 to examine three 9/11 conspiracy theories using two theoretical lenses known as paranoid style theory and cultural sociology (Stempel et al. 2007). The three conspiracy theories are: (1) that the US government assisted or took no action to stop the 9/11 attacks because it wanted the country to go to war in the Middle East; (2) the Pentagon was not struck by an airliner but instead was hit by a cruise missile fired by the military; and (3) the collapse of the Twin Towers in New York was aided by explosives secretly planted in the two buildings. Fifty seven percent of survey respondents believed none of these conspiracy theories, 27% one, 11% two, and 6% accepted all three. The details of these theories are not important to our discussion but the conclusions from this study are worth reporting:

> We found evidence of robust positive associations between belief in conspiracy theories and higher consumption of non-mainstream media (blogs and tabloids), membership in less powerful groups, and personal economic decline. ... The paranoid style theory expects that conspiracy belief will be highest among those least integrated into mainstream social institutions and into the public discourse of the mass media. We found support for both of these hypotheses with the unmarried, those not attending religious services, and the least consumers of a broad range of media associated with at least one of the conspiracy theories. Cultural sociology's unique hypotheses were also supported. Controlling for other media sources, the most legitimate media source (daily newspapers) was negatively associated with two of the three conspiracy theories, and another high legitimacy medium (network TV news) was negatively associated with one conspiracy. Also, the pattern of associations between education levels and the "government assists 9/11" conspiracy fits the claim that the educational differences stem more from the highest status group dismissing conspiracy theories than the most marginalized embracing them. (Stempel et al. 2007)

The second piece of scholarship, by social psychologists Michael Wood and Karen Douglas, considers the intellectual and emotional mindset of conspiracy theorists and their accompanying communication style (Wood and Douglas 2013). In their literature review, they argue that the existing scholarship shows a correlation between people who believe in conspiracy theories and various factors: mistrust of other people and of authorities, feelings of powerlessness and low self-esteem, superstition and belief in the paranormal, a perceived lack of control, a Machiavellian approach to social interaction, and openness to experience.[14] Conspiracy theorists, they assert, feel an obligation to share their views with the public. Their discourse tends to be characterized by an oppositional effect – explicitly arguing against some perceived, official, or widely accepted account of what happened. "Conspiracy theory belief appears to be more of a negative belief than a positive one – it is more concerned with saying what the cause of a condition or event was *not* (i.e., whatever the official explanation is) than with putting forward a specific alternative account."[15]

Wood and Douglas analyzed conspiracist and conventionalist arguments concerning 9/11 that appeared on four major news websites (ABC, CNN, *The Independent*, and *The Daily* Mail) during the final 6 months of 2011 – the time of the 10-year anniversary of the 9/11 attacks. They were particularly interested in studying the 9/11 Truth Movement, a group of people who disputed the mainstream accounts of what happened on 9/11.[16] In addition to those groups and communication media listed in the quotation below, other groups have supported the Truth Movement, including the websites *9/11 Truth* and 911truth.org, the group Scholars for 9/11 Truth and its successor Scholars for 9/11 Truth and Justice, the 9/11 Citizens Watch, and the Hispanics Victim Group.

> [T]he Truth Movement is a well-established community with a substantial intellectual output, including popular books,[17] conference circuits, several sub-organizations such as Architects and Engineers for 9/11 Truth, and at least one peer-reviewed journal, the *Journal of 9/11 Studies*. There is substantial debate within the Truth Movement regarding whether 9/11 was a controlled demolition, a deliberate intelligence failure, or even the result of exotic space-based weaponry (Barber 2008). In short, its body of work is varied, voluminous, and well-developed…" (Wood and Douglas 2013)

Here are the major findings from Wood and Douglas's analysis:

- Approximately twice as many conspiracist comments as conventionalist comments appear on each of these four news sites.
- The overarching attitude of the conspiracists was strong mistrust of authority, and deep skepticism about the way the world is presented to the public.
- Conspiracists were more likely to express mistrust, derogating alternative explanations, especially those coming from officials; whereas conventionalist comments were more likely to express hostility to opposing views.
- Conspiracists and conventionalists alike regarded the phrase "conspiracy theory" as pejorative and were likely to apply it to others but not to themselves.
- People who believed in 9/11 conspiracy theories were more likely to believe in conspiracy theories about other subjects.

---

[14] On the paranormal related to another disaster, the sinking of the Titanic, see Stevenson (1960, 1965).

[15] Wood and Douglas (2013). Emphasis in original. See their literature review for details about this scholarship.

[16] For an ethnography of the 9/11 Truth Movement, see Ellefritz (2014).

[17] See, e.g., Griffin (2004a, b, 2008, 2009, 2011).

- Conspiracy theorists spent most of their time tearing down official positions, whereas conspiracy opponents spent most of their time bolstering their own positions.
- Conventionalists often criticize conspiracy theorists for *anomaly hunting*:

> They imagine that if they can find (broadly defined) anomalies in that data, that would point to another phenomenon at work. They then commit a pair of logical fallacies. First, they confuse unexplained with unexplainable. This leads them to prematurely declare something a true anomaly, without first exhaustively trying to explain it with conventional means. Second, they use the argument from ignorance, saying that because we cannot explain an anomaly that means their specific pet theory must be true. (Wood and Douglas 2013)

The final piece of scholarship we consider here is art historian Peter Ole Pedersen's 2013 analysis of the film *Loose Change 9/11: An American Coup* (Pedersen 2013). This film appeared in several versions during the years 2005–2009 and is one of the most widely circulated accounts of the 9/11 events. It was released on DVD and was also available for streaming over the Internet. It was shown on both American and European television. By 2006, more than four million people had watched it; and it has been downloaded more than 50 million times (Pedersen 2013). The writer and director (Dylan Avery), as well as the producers, had connections to the 9/11 Truth Movement; and the film argues against the official account of the 9/11 terrorist attacks. The film's claims have been widely criticized in the scientific and journalism communities.

Pedersen argues for similarities between *Loose Change* and Oliver Stone's 1991 film, *JFK*, which provides a conspiracy account of the assassination of President Kennedy. Pedersen analyzes Avery's approach, which he claims relies on Avery identifying inconsistencies in the received account, such as the argument that the damage to the Pentagon and the World Trade Center are inconsistent with damage that would have been caused by planes crashing into them. This is entirely in line with Wood and Douglas's discussion of anomaly hunting, mentioned above. Avery alludes – without spelling out complete details – to people in the American government who would gain politically and economically from the 9/11 attacks, and uses this argument to purport to forces in the US government rather than terrorists as being ultimately responsible for the attacks. As Pedersen argues, "During the entire movie, Avery's commentary shifts between an objective rhetoric that accounts for the facts and critical remarks that hint at hidden agendas." The style of presentation involves a wide collection of information set in a documentary style with a "pulsating score" (Pedersen 2013).

The principal means for distributing *Loose Change* has been as a downloaded video from the web. Pedersen explains what he believes makes this genre effective:

> Video sharing websites have precisely manifested themselves as a new series of communication networks, which function as alternatives to already existing top-down media institutions.[18] These new "channels" however appear as hybrids. By way of the archive they draw on our mutual experience from the content of the surrounding film- and TV-culture, while

---

[18] For an account of the impact of 9/11 on media coverage of terrorism, see Powell (2011).

at the same time all this material is integrated in an online-structure, which is defined by the speed of the information transfer. The low-tech characteristics of the clips are a side effect of the migration of the film through the network interface. The video portals appear as cultural institutions on the net and their technological realization paradoxically entails a degradation of the material quality. The image resolution is decided by the fact that we have to be able to stream the content of the archive from one click to the next. … What differentiates these websites from previous frameworks is precisely the active role of the user as contributor of content to the mass media. This type of internet-based cultural institutionalization of filmic material makes it difficult to trace any given truth value back to a single, legitimate source. (Pedersen 2013)

What about conspiracy theories appearing in the alt.folklore.urban (AFU) 9/11 postings?[19] The longest thread of discussion on AFU following the 9/11 attacks was "Rural Reality: Flight 93 Shot Down." It ran from 9/15/2001 to 10/4 and involved 247 posts written by 72 individuals.[20] The thread begins with a discussion of the downing of United Flight 93 in Pennsylvania, the plane that was hijacked but on which the crew and passengers fought back for control. The discussion involves questions about the fact that the federal government has not released a particular cell phone call to the 911 emergency number that allegedly contains the sound of an explosion prior to the crash; about descriptions of white smoke appearing when the plane was still airborne; about contradictory stories of various planes in flight in the vicinity of United 93; about the fact that debris was strewn over a large area; about unusual procedures in nearby radar towers; and about a refusal by the FBI to answer any questions concerning the incident.

Details were compiled by Kevin Keogh, a member of the AFU newsgroup, from approximately twenty major news sources: national and local newspapers, national and local television, and online news sites. Nevertheless, various posters in the thread remained skeptical about these mainstream accounts and questioned whether there was a government cover-up of some sort. The phrase "conspiracy nut" is thrown out, and at one point Keogh writes, "I don't choose to believe fairy tales anymore [sic] than most folks on this newsgroup." Keogh proclaims that it was the government shooting the plane from the sky from a fighter jet and "the tale of the courageous passengers is truly an urban legend". This position is in contrast to another poster (Robert Alston), who responds to Keogh by arguing in favor of a heroic role for the "guys onboard" and stating to Keogh "Oh well. Believe what you want to believe (you will anyway even when it doesn't match reality)." The thread later moves on to a discussion about whether the official account of the collapse of the Twin Towers is feasible.

---

[19] When asked why AFU members would want to address 9/11 rumors when the mandate of the group was to debunk urban legends, long-time AFU member Paul Tomblin (2018) responded: "I think AFU saw it saw it as part of our arena to find the truth behind those, as well as behind the vanishing hitchhiker, the baby train, whatever the famous urban legends were. I mean, it was all part of the spectrum of truth, or finding truth." Tomblin, Paul (2018) (August 2). Oral history conducted by Alexis de Coning for this project.

[20] The entire thread can be read on Google Groups – alt.folklore.urban.

The discussion here – concerning Flight 93 – is not one of the three conspiracy theories discussed by Stempel, Hargrove, and Stempel (Stempel et al. 2007). Nevertheless, if one regards participation in AFU as similar to the higher consumption of non-mainstream media, then the willingness of AFU members to consider alternative explanations to the official one is consistent with these scholars' "evidence of robust positive associations between belief in conspiracy theories and higher consumption of non-mainstream media" – despite the fact that the members of AFU are better educated and presumably in better economic circumstances than the public at large. Certainly, the members of AFU are "least integrated into mainstream social institutions and into the public discourse of the mass media" because of their healthy skepticism for such institutions. The AFU members are more diverse – and presumably more open to evidence and reasoned argument – than the participants in the 9/11 Truth Movement discussed by Wood and Douglas (Wood and Douglas 2013). Nevertheless, the AFU discussion in the thread about Flight 93 is characterized, as Wood and Douglas would predict, both by use of the word 'conspiracist' as a pejorative and by anomaly hunting. The style of discourse on AFU is similar to *Loose Change* in the important role of inconsistencies in the discussions of the shortcomings of official accounts, but AFU is in sharp contrast to the movie, which relies on image and innuendo more than on direct and explicit analysis.

3.4   snopes Working Style

We now turn to analysis of the evaluation of 9/11 rumors and legends on the snopes. com website. This analysis provides a picture of the working style by which snopes evaluates rumors and urban legends. The majority of these analyses are written by Barbara Mikkelson. The others are written by her husband, David Mikkelson, except for one by another snopes' employee, David Emery. Barbara in particular has a distinctive style, and it is generally possible to tell that she is the author without looking at the signature on the evaluation. Almost all of the entries are signed.

While 9/11 is a serious topic and the snopes site treats the topic respectfully, there is also a kind of playfulness in the snopes' evaluations. Just as is the case with the AFU newsgroup, wordplay in particular is valued. The title for a particular rumor evaluation on the website often involves a pun or some other wordplay. For example, one of the rumors concerns the terrorism insurance held by the World Trade Center owners and whether the existence of that insurance was an indicator of foreknowledge of attacks or of owners who did not care whether their building was bombed because they would receive an insurance payout in compensation. The title of this evaluation[21] is Rubble Indemnity, which is a pun on the famous 1944

---

[21] David Emery, "Did a WTC Leaseholder Buy Terrorism Insurance Just Before 9/11? (originally entitled Rubble Indemnity) http://www.snopes.com/wtc-terrorism-insurance/ published 14 September 2016, updated 11 September 2018. All of the URLs referenced in this section were originally accessed in a period of a few days in mid-June 2017 and were revisited in November 2018.

movie *Double Indemnity* concerning insurance fraud. Another example of this playfulness is David Mikkelson's "Alibi Bye",[22] which discusses how the 9/11 disaster caused a man's extra-marital affair to be revealed to his wife.

snopes evaluates rumors using a journalistic approach. The snopes staff is dogged in carrying out library and online research and in tracking down leads by telephone and email with people who might be able to speak with authority about the rumor. Their work is not characterized, however, by using theory as a lens to understand the world more clearly, as a social scientist might do, or by providing a platform from which their analyses might be predictive and testable. Instead, the work is based primarily in a native practical sense. One hallmark of the snopes approach is a healthy dose of skepticism – almost always walking the fine line that would represent being a complete skeptic. Indeed, snopes' evaluations sometimes lead to confirmations, not only to debunkings of 9/11 rumors. In "Angry Muslim Confronts Cashier Over Flag Pin",[23] Barbara Mikkelson reflects on the practicalities of snopes' fact-checking methodology: "When confirming evidence is lacking, one should strive to remain skeptical of what are presented as real-life accounts that state in narrative form things people are predisposed to believe, especially those tales wherein wrongdoers get their comeuppance through being told off by others."[24]

One advantage that the snopes staff applies regularly is their broad knowledge of 9/11 events and rumors, which enables them to uncover patterns among rumors over time and explain how the particular rumor being evaluated fits into a larger pattern. For example, in "Warning from Terrorist," Barbara Mikkelson traces the emergence and re-emergence of the Grateful Terrorist rumor, tying each new appearance to a recent event that may have stimulated a re-emergence of the rumor – not only the 9/11 attacks in 2001, but also the death of bin Laden in 2011, a hostage situation in Sydney, Australia in 2014, and a terrorist attack in Paris in 2015.[25]

Similarly, the employees at snopes have a general knowledge of folklore that enables them to set particular 9/11 rumors into a historical context. In the evaluation "Women Killed by Poisoned Perfume Samples?",[26] David Mikkelson points to the similarities of this particular 9/11 rumor and two earlier rumors that predated the 9/11 events:

> This baseless bit of scarelore appeared to be a combination of two older, equally unfounded pieces of the same genre: the perfume robbers tale (women in parking lots lured into sniffing cut-rate perfume lose consciousness and are robbed while they're out) and the Klingerman virus scare (blue virus-laden sponges mailed in envelopes marked "A gift for

---

[22] David Mikkelson, September 11 Adultery Revelation (originally published as Alibi Bye) http://www.snopes.com/rumors/adultery.asp published 5 October 2001, updated 11 September 2018.

[23] David Mikkelson original (25 March 2015), updated by Barbara Mikkelson (9 August 2015), "Angry Muslim Confronts Cashier Over Flag Pin" http://www.snopes.com/rumors/lapelpin.asp

[24] Mikkelson and Mikkelson, Angry Muslim Confronts Cashier Over Flag Pin.

[25] Barbara Mikkelson, Warning From Terrorist, http://www.snopes.com/rumors/warning.asp originally published 8 January 2006, update 11 December 2015.

[26] David Mikkelson, Women Killed By Poisoned Perfume Samples?, originally published 3 November 2001, updated 4 June 2017, http://www.snopes.com/rumors/perfume.asp

you from the Klingerman Foundation" have caused 23 deaths). But lore moves forward with the times, so this newer caution incorporated "terrorists" (presumably Middle Eastern) into the mix.[27]

The snopes staff is well aware of the multiplying effects that the Internet can have in spreading rumors. "Billy Graham's Daughter's Speech"[28] concerns the rumor that Graham's daughter appeared on television, where she chastised the American public for turning to God only in times of crisis. In this rumor evaluation, an anonymous member of the snopes staff directly addresses the multiplying effect of the Internet on communication: "As many others have pointed out, one of the pitfalls of the Internet as a means of communications is that it can be used to spread misinformation as rapidly as accurate information, but all too often the former is set loose to spread far and wide, and correction comes too late, if at all."[29]

Many of the evaluations draw in a non-scholarly way on the social psychology of terrorism. In "Alibi Bye" David Mikkelson reflects on sick humor as a coping mechanism in the face of catastrophe, except that some events, such as 9/11, are so horrific that at least for a time there is no room for humor, even sick humor[30]:

> Humor almost disappeared from American culture in the period immediately following the tragedies of 11 September 2001. For more than two weeks nary a joke was to be heard throughout the land, let alone any of the to-be-expected outpourings of sick humor that often chase close upon the heels of horrific events. The Challenger explosion, the death of Princess Diana: those were almost immediately the subject of gallows humor offerings, some of which were so appallingly tasteless as to provoke a smile in even those most reluctant to laugh. This time things were different: the terrorist attacks and the staggering loss of human life on that beautiful September morn silenced the laughter.[31]

In a similar vein, in "Bodies in Airplane Seats Found"[32] Barbara Mikkelson discussed the psychological difficulties in coming to terms with devastation on such a massive scale, and how people might better understand the overall tragedy by focusing on the tragedies that faced particular individuals or particular small groups of people:

> The aftermath of the September 11 terrorist attacks was fraught with baseless rumors about grisly discoveries made at and near Ground Zero in New York. ... The horror we felt (and still feel) found an outlet in those tales, so we passed them along. Each gruesome piece of scuttlebutt was an attempt to put into words our gropings towards a better understanding of the butchery that had taken place. Sixty-five dead on Flight 175 was beyond our comprehension – we understood the number, but could not take in the loss in terms of sixty-five individual tragedies. But one trampled baby we could understand, as we could one

---

[27] David Mikkelson, Women Killed By Poisoned Perfume Samples?

[28] David Mikkelson, Billy Graham's Daughter's Speech, originally published 3 October 2001, updated 9 March 2018 (http://www.snopes.com/rumors/wheregod.asp)

[29] David Mikkelson, Billy Graham's Daughter's Speech

[30] For a collection of 9/11 jokes, see Ooze.com (2006). For reflections on the value of dark humor, see Gournelos and Greene (2011); Brottman (2012); Kuipers (2005); and Ellis (2002). On 9/11 and the popular culture of comics, see Dittmer (2005); and Peitz (2013).

[31] David Mikkelson, September 11 Adultery Revelation.

[32] Barbara Mikkelson, Bodies in Airplane Seats Found, published 21 April 2008, http://www.snopes.com/rumors/strapped.asp

tragic policeman who put a gun to his head rather than wait for a rescue that would never come. Those mental images haunted us, but they also helped us put into context horror that existed on so grand a scale it would otherwise make little sense.[33]

Several of the rumor evaluations addressed coping mechanisms used by ordinary American citizens after the 9/11 attacks. One approach snopes identified was to emasculate or otherwise diminish the enemy. The material in the section of this chapter on photoshopped images discusses how digitally modified images intended to diminish Osama bin Laden serve this purpose. In "Osama bin Laden Kidney Disease Rumor",[34] Barbara Mikkelson describes how rumors about bin Laden functioned psychologically to diminish his power over us. This, she argued, was a common theme, not true of bin Laden only but of other enemies now and in the past:

> In the world of gossip, scuttlebutt that trivializes a hated enemy by portraying him as less powerful and more piteous often proves highly popular because it helps reduce the perceived threat that person represents to a more manageable level. Such rumbles are routinely kited during times of conflict, and often prove to have little more to them than mere wishful thinking on the part of those looking for reassurance.[35]

If speculation about bin Laden having a serious, perhaps fatal disease played this positive psychological function for American citizens, stories about his body washing ashore 10 years after the 9/11 bombings may have – counterintuitively – had a more negative connotation. In "Osama bin Laden's Body Washes Ashore,[36] "Barbara Mikkelson puts forward several possible uncomfortable symbolic meanings of bin Laden's body washing ashore:

> Rumors of bin Laden's washing up somewhere were almost to be expected. In one sense, they are a way of voicing the uncomfortable reality that the threat he represented didn't die with him, but instead is only momentarily out of sight and sure to return before long. The al Qaeda leader may be dead, but al Qaeda itself is still operative, as are other terrorist organizations. Bin Laden had also become a cult figure in the Middle East, thus stories about his body's coming back are a way for those who regarded him as a symbol of resistance to American imperialism to express that the man wasn't that easily gotten rid of, nor will be what he stood for. Last, from a purely Western point of view, such yarns could be interpreted as a symbol that even the sea didn't want him.[37]

snopes explains that that the American public wanted to find ways in which they could participate personally in the fight against the terrorists. Barbara Mikkelson argues in "Osama bin Laden Owns Snapple Rumor"[38] that, since the ordinary American could not be on the ground fighting in Afghanistan, this is why the

---

[33] Barbara Mikkelson, Bodies in Airplane Seats Found.

[34] Barbara Mikkelson, Osama bin Laden Kidney Disease Rumor, 1 May 2011, http://www.snopes.com/rumors/kidney.asp

[35] Barbara Mikkelson, Osama bin Laden Kidney Disease Rumor.

[36] Barbara Mikkelson, Osama Bin Laden's Body Washes Ashore, updated 6 May 2011, http://www.snopes.com/rumors/ashore.asp

[37] Barbara Mikkelson, Osama Bin Laden's Body Washes Ashore.

[38] Barbara Mikkelson, Osama bin Laden Owns Snapple Rumor, updated 21 April 2008. http://www.snopes.com/rumors/snapple.asp

American public was so willing to participate in a boycott of the Snapple beverage company. There was a widespread rumor that Osama owned the distribution rights to Snapple in the Middle East, and so, by boycotting the company, one could reduce bin Laden's ability to support terrorism.[39]

snopes also found that, in their search for appropriate reactions to the 9/11 terrorist attacks, some Americans found solace in simplistic solutions such as blanket carpet bombing of Afghanistan or wholesale denial of immigration from Arab states. Another approach, snopes noted, was to find ways to turn Muslim religious beliefs and practices against the terrorists. As David Mikkelson explains in "Pershing the Thought" [another pun!], according to a false rumor that circulated after 9/11, General J.J. Pershing had quelled a Muslim insurgence in the Philippines in the early twentieth century by burying the insurgents killed by American troops with pigs, a repugnant act for the insurgents because of the Muslim prohibition from eating pork. A version of this story was apparently told by Donald Trump when he was on the campaign trail during the 2016 presidential election.[40]

> Nonetheless, the desire for simplistic solutions to complex problems has spawned several widely-circulated notions that seek to transform a fight against terrorism to the easily-manageable level of a horror film or a comic strip. One popular notion is the concept that a pig is to a Muslim as a crucifix is to a vampire: simply arm yourself with a porker, and you can use it to render even the most fanatical terrorist helpless, sending him cowering in fear lest he come into contact with anything porcine. Such notions reduce an extremely widespread and diverse religion, and the people who follow it, to a monolithic entity with a single set of beliefs and rules to which everyone adheres. … Also implicit in this type of reasoning is the notion that "terrorist," "Muslim terrorist," "fanatical Muslim" and "devout Muslim" are all synonymous. They aren't…[41]

Another kind of simplistic reaction identified by snopes was a variety of total, fervent patriotism that brooked no disagreement and no qualification. In "University Peaceniks,"[42] David Mikkelson evaluated an urban legend that circulated in various forms online about the confrontation between the heroic everyday Joe and the liberal, anti-war university professor:

> The post-Sept. 11 climate of patriotic fervor encouraged the spread of stories such as this one. Through them, folks gain a vicarious sense of having participated in defending their country (even if that "defending" amounts to putting the run on loudmouth eggheads at the local Best Western). In times of war, liberal attitudes are less tolerated, and old wounds (e.g., the Hanoi Jane chapter of the Vietnam War) are again scratched open. Less complex world views are adopted, and many draw strength from a return to the jingoism of "Love it

---

[39] As Barbara Mikkelson explains, the Middle Eastern distribution rights to Snapple were owned at the time of 9/11 by the Binladen Group, which is a part of Osama's extremely large family, but he apparently had no connection to this company; and, in any event, Cadbury Schweppes, which owns Snapple, severed its relationship with the Binladen Group soon after 9/11.

[40] David Mikkelson, General Pershing on How to Stop Islamic Terrorists (originally appearing as "Pershing the Thought", originally published 31 October 2001, updated 17 August 2017, http://www.snopes.com/rumors/pershing.asp

[41] David Mikkelson, General Pershing on How to Stop Islamic Terrorists.

[42] David Mikkelson, University Peaceniks, published 9 March 2008, http://www.snopes.com/rumors/egghead.asp

or leave it" and "If you're not with us, you're against us." Overly liberal attitudes are lampooned in this piece through the invocation of diametrically opposed stereotypes: the armed services veteran who has sacrificed much for his country and the stay-home college boor who is long on opinion but short on real-life contribution to the greater good. But of course the populace (in this case, the folks in the restaurant) rise up to support the patriots, thus aligning themselves on the side of right.[43]

In several of the rumor evaluations, Barbara Mikkelson made interesting comparisons between the war on terrorism and both the Second World War and the Cold War. During the Second World War, there were various stories of Nazis roaming American streets, signaling to Axis submarines and scouting out munitions factories, unrecognizable to ordinary Americans until they revealed their hands. Mikkelson explains in "Osama bin Laden in Utah"[44] the functional meaning of such rumors:

> Sightings of the dread enemy walking among the populace he menaces are a common form of wartime scuttlebutt and serve as an expression of the reality of the danger he represents. A rumor of that nature works to keep the minds of those on the homefront focused on whom they are fighting and how great an evil he embodies; they make the threat he represents seem more real to those who might otherwise feel somewhat distanced from the conflict underway in another theater.[45]

However, Barbara Mikkelson also finds differences between the Arab terrorists and the Nazis. This modern terrorism she regarded as less honorable (for it is not state fighting state – although it is hard to argue that what the Nazis did to the Jews in concentration camps is in any way more honorable) and perhaps more terrifying because it was harder to control:

> In wartime and during the ramp-up towards war, leaders of opposing forces come to loom in our minds as demonic figures that are the embodiment of evil. Because we like simple solutions to complex problems, the notion that if something happened to those men, the enemy forces they lead would fold their tents and sneak away into the night is seductive, hence the much-loved aphorism about killing a snake by chopping off its head. Consequently, rumors about the demise of enemy leaders are typical fare during times of political tension. Throughout World War II, persistent whispers about the secret deaths of Hitler, Mussolini, and Tojo abounded, with similar buzz about Stalin surfacing afterwards as events subsequently served to make him appear to be the next great threat to world peace.... Whereas a Hitler or a Tojo might have needed to die to give the average man on the street a sense of assurance that the war could be won, perhaps no similarly cataclysmic event need be provided in a conflict where the enemy forces aren't viewed as a unified, national fighting force. But the difference could just as easily be attributable to death not being regarded as a great enough punishment for Osama bin Laden. At the heart of the "dead leader" rumors of World War II lay the average person's desire to see the threat to his personal safety lifted – though there was certainly room in the process for gleeful thoughts of retribution, compared to the war's being brought swiftly to an end, the enemy's leader being punished was not nearly as important an outcome. In the eyes of a Western populace unaccustomed to acts of

---

[43] David Mikkelson, University Peaceniks.

[44] Barbara Mikkelson, Osama bin Laden in Utah, last updated 21 April 2008, http://www.snopes.com/rumors/utah.asp

[45] Barbara Mikkelson, Osama bin Laden in Utah.

terrorism, the 2001 attacks on America were not instances of ordinary warfare. Their mastermind was viewed as a murderer, not as a military leader, so the desire to see him humbled and brought in chains to face judgment ran at least apace with the yearning for a return to the pre-2001 days of unquestioned safety. Death may well have been thought too good for him.[46]

The evaluation of two rumors that began circulating in late 2001 concerned the ordinary American's feeling of isolation and not knowing what was going on in the rest of the world. How did Europe or Asia feel about what had gone on in the United States on 9/11? What was daily life like for the American troops in Afghanistan as they sought out bin Laden and his terrorist network? Some Americans did not believe they were getting the full picture from CNN and other mainstream television sources. The first of these rumors concerned news reporting about 9/11 from other parts of the world. snopes noted how eager and pleased Americans were to learn about the mostly pro-US stance taken in these articles.[47] The second rumor concerned a letter, seasoned with salty language, that began to be circulated in late 2011. It was allegedly written by a Marine on site in Afghanistan concerning the conditions under which the US troops lived and worked. The letter was read on some commercial radio broadcasts and reported as true, although it was eventually shown that the letter was made up.[48]

In "Dunkin Donuts 'Celebrating Employees' Rumor,"[49] Barbara Mikkelson expressed particular concern about the harm caused by mercantile rumors. Any company that people believed either took actions that abetted terrorists or employed workers who supported terrorists or celebrated their actions were ready targets for economic boycott, or for more serious and violent actions. Rumors of this sort spread rapidly online. While there was no basis in truth in almost all cases of these rumors, they could be extremely harmful to businesses both large and small. Moreover, there was only a limited amount that the targeted companies could do to dispel the rumor, and essentially nothing they could do to avoid becoming the target of this kind of rumor in the future.

> The rumors about employees of Arab extraction celebrating upon receiving news of the attacks on the World Trade Center and the Pentagon have been attached to numerous businesses beside Dunkin' Donuts. What needs be kept in mind is that these kinds of rumors are not specific to any one company, therefore any thoughts of the "Where there's smoke, there's fire" variety should immediately be dismissed. They're baseless rumors, and whom they're aimed at has nothing to do with the named party's having done anything to bring it upon itself. … Large chains aren't the only commercial entities tarred with this undeserved brush; numerous small firms had versions of the same slander applied to them. According to breathless rumor spread willy-nilly, Arabs have been caught in the act of celebrating the

---

[46] Barbara Mikkelson, Osama bin Laden Captured, last updated 1 May 2011, http://www.snopes.com/rumors/captured.asp

[47] David Mikkelson, "An Ode to America", originally published 12 October 2001, updated 3 August 2018, http://www.snopes.com/rumors/soapbox/nistorescu.asp

[48] Barbara Mikkelson, "SaucyJack Letter", 8 March 2008, http://www.snopes.com/rumors/freezing.asp

[49] Barbara Mikkelson, Dunkin Donuts 'Celebrating Employees' Rumor, updated 6 December 2005, http://www.snopes.com/rumors/dunkin.asp

strike against the twin towers and the Pentagon in bagel shops, restaurants, stores, and coffee houses – anywhere customers could conceivably have witnessed such outpourings. False rumors like these run on very fast legs indeed, and spontaneous boycotts sprang up in their wake. These boycotts have done irreparable harm to the many innocent businesses swept up by this wave of misinformation. Combatting a rumor is rarely an easy task. The very nature of gossip almost guarantees that a tale's originator will not be found, nor will any of its early disseminators. By the time a false charge has grown large enough for its effect to be noticed (or to even to fall upon the ears of those defamed by it), the ones who started the smear are long gone, while their creation spreads outwards in exponential fashion. Those to whom falls the unhappy task of quelling the harmful rumors that have attached to their firms at least have a bit of a chance at getting to the source when what was said is distributed via e-mail. Gossip spread in written form is a tad easier to track than gossip passed behind hands, but tracing it is still a herculean task, and it never comes with a guarantee of success.[50]

This Celebrating Arabs story is one of the most popular memes in the 9/11 rumor mill. snopes also has paid close attention to two other popular memes, which are closely tied to one another: foreknowledge legends and Grateful Terrorist legends. Both involve some group of people having prior knowledge of some disaster that is about to happen, and the latter type of legend goes beyond that to have the person with the foreknowledge warning an innocent American because of some kindness that has been paid to them. Rubble Indemnity, described earlier in this chapter, is one example of foreknowledge, but so are the stories about employees who called in sick on 9/11, the absence of taxis around the World Trade Center that fateful morning, or of those people who used foreknowledge of the event to make a fortune by "shorting" stock in airline companies.[51] Of the taxi rumor, Barbara Mikkelson wrote:

> The rumor was a way of putting into words a chilling realization America was fighting to come to terms with: Those who had perpetrated the attacks had lived among their victims without raising suspicion, and others of their kind were still out there, quietly biding their time and waiting for their turn to strike. Every male of Arab descent was now suspect, and the presumed loyalty of immigrants from the Middle East was being weighed by all. People were left to wonder whom they could trust. Whom they could really trust. Was country of origin and its culture more important to émigrés from the Middle East than allegiance to the new country they had chosen to make their home? This rumor put into words America's fear that it did, because at its heart it asserted that even those who weren't directly involved in the attacks must have had knowledge of what was coming but did not see fit to warn others who weren't of their blood. When push came to shove, said the rumor, their loyalty hadn't been to the country that had opened its arms to them; it had been to the murderers from back home.[52]

Foreknowledge legends are closely related to serendipity legends. In the 9/11 serendipity legends, for example, a man did not show up at work in the World Trade Center the morning of September 11 because he had a meeting with his daughter's

---

[50] Barbara Mikkelson, Dunkin Donuts 'Celebrating Employees' Rumor.

[51] For an empirical study of possible insider trading on the stock market just prior to the 9/11 attacks, see Wong et al. (2011). They argue that there was unusual trading at this time.

[52] Barbara Mikkelson, No Taxis at WTC, last updated 21 April 2008, http://www.snopes.com/rumors/taxi.asp

guidance counselor and was going to go to work late. The man had no foreknowl-edge of the terrorist activities, but the result was the same – he was safely outside of the World Trade Center at the time of the plane crashes. But, unlike cases of fore-knowledge, he had no reason related to the terrorist attacks not to have gone to work.

Barbara Mikkkelson explains in "Oct. 26 Terrorist Attack on Los Angeles Rumor" that the other popular meme, the Grateful Terrorist legend, is highly implausible:

> ...terrorists do not telegraph their plans to those who could potentially interfere with the success of their operations. Those who have devoted their lives to bringing about the downfall of a particular government or country will not jeopardize their thirsted-for goal merely to spare a stranger, a daughter of a friend, or even a wife. The terrorists who attacked America on September 11 boarded planes and looked into the faces of those they would kill, and it meant nothing to them.[53]

However, this type of rumor, she noted, remains popular because it serves an important psychological function, to give us control over a life-threatening situation:

> The blood-curdling reality of terrorism – even outstripping the death, destruction, mayhem, and unspeakable tragedy in terms of individual lives lost – is its unpredictability. No one knows when or where the next strike will be. How comforting it thus is to entertain such tales that tell you the "when" and the "where" and thus return the issue of your own safety back into your own hands rather than leaving it to the whim of the homicidal. We like such rumors because they reduce the chilling reality of "Anyone could be a target at any time" down to a matter of "Don't be in Los Angeles on the 26th." One is unforeseeable (and thus unstoppable), while the other is something we can exert a measure of control over.[54]

We close this section with a discussion of one final evaluation by Barbara Mikkelson, of a rumor entitled "Post 9/11 Baby Boom."[55] It had been widely believed that a side effect of the 9/11 terrorist attacks would be a major jump in births 9 months later. In fact, maternity wards across the nation, but especially in New York City, geared up for this increase but it never materialized. In explaining why, Mikkelson provides a thoughtful and nuanced analysis of the human responses to 9/11:

> The September 11 attack on America caused many to take a long, hard look at their lives. For some, that re-evaluation led them to realize that the time to start a family was now. September 11 was a major shock to their systems, jolting them from a state of somnambulistic "We'll get around to it someday; we've all the time in the world" complacency into the wide-awake appreciation that life was both precious and uncertain, and that the perfect tomorrow they were waiting for might never come. There were those who in the wake of the attacks quit putting off what they'd previously been half-hearted about, and they started families, got married, or recommitted to their existing marriages. Times of crisis cause us all to look into our hearts and see if we're truly concentrating on what really matters rather

[53] Barbara Mikkkelson, Oct. 26 Terrorist Attack on Los Angeles Rumor, updated 7 April 2008, http://www.snopes.com/rumors/oct26.asp.

[54] Barbara Mikkkelson, Oct. 26 Terrorist Attack on Los Angeles Rumor.

[55] Barbara Mikkelson, Post 9/11 Baby Boom, updated 7 March 2008, http://www.snopes.com/rumors/babyboom.asp

than stumbling along in unthinking routines. But that realization is a double-edged sword, a fact the pundits who predicted a startling upsweep in the number of weddings and births failed to take into account. For every gal who decided that yes, this man was the one, and it was time to take the walk down the aisle, there was just as likely another woman who realized the fellow she'd been considering marrying wasn't the man she wanted to spend the rest of her life with. For every fellow who decided to make peace with his wife, there might just as well have been another who realized he was throwing away his life on the wrong woman. And just as there were couples who deemed that no matter how the world was faring, now was the time to begin a family, there were others who experienced second thoughts about having children and opted to postpone the project until the world had steadied itself. Catastrophe, be it a personal tragedy or a communal horror, often serves as a wake-up call to those who've been letting their lives slip through their fingers, but how individuals will choose to react to such a clarion call is unpredictable. Yes, important decisions may be spurred by horrifying events, but people don't simply rush out *en masse* and all follow the same course of action, because they all have different lives and viewpoints.[56]

3.5   Additional Treatments of 9/11 Urban Legends

snopes was not the only group tracking 9/11 urban legends. The mass communication scholar Charles Hays has compared the early reactions to 9/11 on two Usenet newsgroups, soc.culture.scottish[57] (SCS) and rec.motorcycles.harley[58] (RMH) with a goal of understanding how they tried to "repair" their community identity in the face of these horrific acts. It turns out that the ways in which these two Usenet groups handled the tragedy were significantly different (Hays 2011).

The 9/11 news dominated discussion on RMH on the day of the 9/11 terrorist attacks. For a while, as news was still drifting in, there was reluctance to take a position. Comparisons were drawn to the Japanese attack on Pearl Harbor, and there was discussion of how the US government should respond to the 9/11 attacks. However, soon many members were calling for a violent military response to what they perceived to be a threat to the American way of life; those who were calling for restraint were shouted down.

The SCS newsgroup used Usenet as a way to communicate news about the terrorist attacks in the early hours after they occurred since other news media were slow in providing details and analysis. Many of the early messages were expressions of concern about people they knew who might have been caught in the attacks. The group worried about the mental stability of President Bush and how he would react to the attacks. SMS was more tolerant than RMH of diverging viewpoints about what the response to the attacks should be, and there was an effort to maintain a sense of community within SMS. Consensus built more slowly in SCS than in

---

[56] Barbara Mikkelson, Post 9/11 Baby Boom.

[57] soc.culture.scottish (2001). Discussion thread: "New York Tragedy." Archived at http://groups.google.com

[58] rec.motorcycles.harley (2001). Discussion thread: "OT: Opinions on what US should do?" http://groups.google.com

RMH, and the consensus that did build was for a nuanced international response instead of a unilateral one by the American military. As Hays reported of SCS, the attitude was that "surgery needs a scalpel not a shotgun" (Hays 2011).

But how did alt.folklore.urban (AFU), the Usenet newsgroup focused most closely on urban legends, respond to 9/11? From a single day's postings – the method used by Hays – one might get a good measure of the initial reactions of the community to the terrorist attacks, but it is hard to gain a complete picture of how the AFU newsgroup handled 9/11. So instead of analyzing traffic for a single day, we analyze traffic for the first 30 days following the 9/11 attacks (threads ending 9/11 through 10/10/2001). During that time there were 353 threads, 206 (58%) of which concerned 9/11. (The threads are listed at https://scholar.colorado.edu/infosci_facpapers/4/. It is often but not always possible to understand the topic of the thread from the thread title.)

Some of the earliest threads concerned news about the attacks, expressions of grief, and prayers for the victims. However, only 2 days later, on 9/13, members began to examine urban legends related to the 9/11 attacks, such as a thread about whether the plumes of smoke from the World Trade Center could be seen from space and the first of many instances questioning whether someone had foreknowledge of the attacks. In a thread that began on 9/11 and ended on 9/15, entitled "and so the folklore starts", one could see that the newsgroup anticipated there would be significant folklore associated with this tragedy. Discussion in that thread included the Nostradamus prophecy, several other suspicious plane crashes and emergency landings, and discussion of the over-inflation on the death toll by the media in New York City as September 11 went on. Not surprisingly, many of the topics covered on AFU were the same as those covered on snopes, e.g. Celebrating Arabs stories, a man safely riding down the rubble in the World Trade Center, stories about bin Laden sightings, and examples of numerology and pareidol. However, a few of the threads concern 9/11 urban legends that did not appear on snopes, for example one about hijacked Greyhound buses. In less than a week's time, the 9/11 discussions on AFU were overwhelmingly about critical assessment of 9/11 urban legends.

Some of the postings appearing on AFU were unlike those on either SCS or RMH, including a meta-analysis of urban legends related to 9/11 (e.g. a thread about whether it was ever possible for a "friend of a friend" story to be true) and postings that tracked news of snopes and other members of AFU who were being covered by the media as they debunked 9/11 urban legends.

Another of the threads, "Apopolgy",[59] reflected on the acceptable standards of behavior on AFU. The acronym BoP is used as shorthand on AFU to express the prohibition from making political comments (Ban on Politics) on the newsgroup.

---

[59] Alt.folklore.urban, archived at Google Groups, https://groups.google.com/forum/#!forum/alt.folklore.urban, the thread is entitled "Apology", ending September 15, 2001. In subsequent endnotes, we simply give the name and the month and day of the thread (All are from year 2001.). There are search tools on Google Groups that help one to find the appropriate thread.

Michael begins this thread by apologizing for breaking the BoP rule, but also explaining his actions by the exceptional circumstance of the terrorist attacks:

> I'm sorry if this weeks events has forced/allowed me to cross the line to politics (and bad temper) [sic]. That was not my intention.
>
> Part of it came from my emotional state. Part of it was in direct response to posts made to my posts that I felt required a some politically oriented defense. Understand, I have tried to practice self restraint. … much of what I've *thought* has not made it to this group. I have taken that elsewhere.
>
> I understand the ban, but I tend to be defensive – you saw the worst during this tragedy. Admittedly there is a fine line between some of these topics and the political content. I failed to draw that line. Or, as an excuse, I failed to see the line.[60]

Another poster (calling him/herself +) in the thread, is incredulous about Michael's apology and flames another member of the newsgroup who actively enforces the BoP rule:

> Let me get this straight, you're apologizing for expressing yourself in a public forum where you can post whatever you like?
>
> You're going to let a net abusing bully like Simon Stalin chase you off the group because by his reckoning, you're doing something wrong?
>
> They'll have to pry the keyboard from my cold and clammy dead hands before I let some fuckhead like Simon tell me that I can't express my patriotism, my uncle died defending this country from people that would take away your right to free speech.
>
> Maybe Simon is an [sic] member of the Moro Islamic Liberation Front, anyone check his ancestry? He sure looks dark and swarthy to me.[61]

This discussion of rules and norms on AFU continues in another thread[62] in which one member, Ed Kaulakis, posts a statement about the attacks and Simon criticizes Ed for this posting being off-topic and hence inappropriate for AFU, as the two excerpts below show.

> The enemies of the United States proved on Sep 11 that they are clever, determined, well-organized, deadly, and courageous unto death. Let us give them that respect; it will prevent foolish miscalcutions [sic]. And then let's treat them as enemies. (Ed Kaulakis)

> With the best of intentions I don't see what that has to do with alt.folklore.urban. Could you please take it elsewhere? (Simon Slavin)[63]

One person complained to Simon that he was being too heavy-handed, given the circumstances:

> Simon, I love you like a brother, but could you perhaps ease up a bit for the nonce? In the interests of the occasion? A bit of clutter in the NG for the time being is not a big deal right now. There's a good chap. (Tom Cikoski)[64]

---

[60] "Apology", thread completed September 15.

[61] "Apology", thread completed September 15.

[62] "Enemies", thread completed 9/17.

[63] "Enemies", thread completed 9/17.

[64] "Enemies", thread completed 9/17.

But another member argued that there was a good reason to enforce the BoP rule:

> But it seems to me that your intent – that we ought care tenderly for our fellow afuisti in this trying time – is best served by refraining from stoking the incredible fury some of us are feeling right now. And to that end, the BoP is our friend. (Chris Clarke)[65]

This discussion of the ground rules for AFU continues for 31 posts, involving 18 people, with both sides being argued. In yet another thread, "Proposal for the Group",[66] Mark D. Lew calls for a return to normalcy, as Warren G. Harding would say, online as well as in the physical world:

> On Monday, the New York Stock Exchange will reopen after being closed for several days. On Monday, Major League Baseball will resume play after several days suspension. These interruptions, and various others, were understandable in the wake of the tragedy on Tuesday. However, the time comes when we wish to attempt to return to "normal" life.
>
> In AFU, the BoP (along with various other Bo's) has been partly suspended. This too is understandable, given the circumstances, but it will soon be time to return to normal.
>
> I would like to see the BoP up and running again by Monday morning. Can we perhaps agree on this?[67]

Another post, from the thread "BOJJ, etc",[68] written by an American citizen (Angie SCHULTZ) who was in Australia at the time of the terrorist attacks, discusses the value that AFU offers as a virtual community:

> I find Usenet a surprising comfort under these circumstances. I can't really wander blindly around bonding with my neighbors, who naturally treat this as a distant evil. I want to come home, not that I could help much (especially since "home" would be California). I can't even charge down and attempt to give blood (and be turned away, as usual). So it's good to "hear" other people worrying, speculating, dithering, ranting – just like I could do with them if I were there.[69]

How many postings were there in these threads and how people participated in them? The left half of Table 3.2 shows the frequency distribution of the postings in a given thread on AFU related to 9/11 in the month after the attacks. You can see that the mode is a single post in a thread, the maximum number of posting in any of these threads is 247, and roughly half of the threads contain 9 or fewer posts. The right half of the table shows the frequency distribution of the number of unique people posting in the thread. The mode is again 1, the maximum number of people posting is 80, and well over half of the threads involve fewer than 10 people.

We do not have space to carry out a complete analysis of all the 9/11 threads on AFU, but we discuss a few select threads, especially those that had large numbers of postings. Above, we have already discussed the longest thread on AFU concerning the 9/11 attacks, "Rural Reality: Flight 93 Shot Down,"[70] which raises the question of a government conspiracy in the official accounts of both the downing of United Flight 93 in Pennsylvania and the collapse of the Twin Towers in New York.

---

[65] "Enemies", thread completed 9/17.
[66] "Proposal for the Group", thread completed 9/17.
[67] "Proposal for the Group", thread completed 9/17.
[68] "BOJJ, etc", thread completed 9/29.
[69] "BOJJ, etc", thread completed 9/29.
[70] "Rural Reality: Flight 93 Shot Down," thread completed 10/4.

Table 3.2 Frequency distribution of the postings and unique people posting per thread on AFU related to 9/11 in the month after the attacks

Number of postings in thread	Frequency	Number of unique people posting in thread	
1	34	1	34
2–4	37	2–4	45
5–9	36	5–9	39
10–25	27	10–25	30
26–50	21	26–50	17
51–100	8	51–100	5
101–200	5	101–200	0
>200	1	>200	0
Max = 247		Max = 80	

Source: Calculated from the data in https://scholar.colorado.edu/infosci_facpapers/4/

The second longest thread on AFU concerning 9/11 was "Reflections on a date which will live in infamy",[71] which included 175 postings from 69 people and lasted from 9/12 to 9/24/2001. Mostly this is a chance for people to talk, explore various dimensions of the attacks, and try to make sense of them. One small portion of this discussion came back to the conspiracy theories that were front and center in the thread "Rural Reality".[72] On this topic, Andy Walton wrote sarcastically: "There's a large body of conspiracy theory based on initial reports that didn't show up later. If something shows up in the first minutes or hours that isn't mentioned weeks later, it must be the subject of a coverup."[73]

The other lengthy AFU threads on 9/11 concern topics already discussed in the section on snopes's discussion of 9/11 rumors and need little further discussion: numerology ("Eleven Eleven," ending 10/3, with 131 postings and 67 participants),[74] self-censorship on radio musical offerings ("Clear Channel will protect your tender sensibilities," thread ending 10/4/2001 with 116 postings and 57 participants),[75] worrisome events that might be connected to terrorism ("Stolen Verizon Trucks: Latest hoax/legend," thread ended 10/6/2001 with 115 postings and 43 participants),[76] and prophecies ("Nostradamus Folklore," thread ended 9/24/2001 with 82 postings and 40 participants).[77]

---

[71] "Reflections on a date which will live in infamy", thread ending 9/24.

[72] "Rural Reality," thread completed 10/4.

[73] "Rural Reality," thread completed 10/4.

[74] "Eleven Eleven," thread completed 10/3.

[75] "Clear Channel will protect your tender sensibilities," thread completed 10/4.

[76] "Stolen Verizon Trucks: Latest hoax/legend," thread completed 10/6.

[77] "Nostradamus Folklore," thread completed 9/24.

3.6   Harms to Businesses by Mercantile Rumors

In the aftermath of 9/11, a number of businesses were harmed by unsubstantiated (and generally false) rumors spread on the Internet about actions by their employees that were regarded as unpatriotic. Consider the following example, collected by snopes.com in 2001:

> Attention all Americans: Boycott Dunkin Donuts!!
>
> In Cedar Grove, NJ, a customer saw the owner of a Dunkin Donuts store burn the U.S. flag. In another Dunkin Donuts store in Little Falls, a customer saw a U.S. flag on the floor with Arabic writing all over it. In Wayne, NJ the employees of Arabic background were cheering behind the counter when they heard about the attacks. A customer threw his coffee at them and phoned the police.
>
> We are starting a nationwide boycott of all Dunkin Donuts. Please make sure this gets passed on to all fellow Americans during this time of tragedy. We Americans need to stick together and make these horrible people understand what country they are living in and how good they used to have it when we supported them. Numerous fastfood companies are at Ground Zero, giving away free food to volunteers. Where is Dunkin Donuts in all of this? Boycott Dunkin Donuts! Pass it on.[78]

A similar urban legend spread after 9/11 about a Budweiser delivery person, who upon allegedly seeing Arabs at a 7-11 store celebrating the 9/11 attacks decided to pull all Budweiser products from the store.[79]

A different kind of harm to business was created through the Grateful Terrorist legend. As the legend goes, in gratitude for a kindness from an American citizen, an Arab man tells the citizen not to drink Coca-Cola products (presumably because the Arab has foreknowledge that terrorists were adulterating the products). This legend spread so widely that Coca-Cola had to release a press notice in 2002 that the rumor is false and that Coca-Cola products are safe. Apparently, similar rumors about grateful foreigners revealing adulterated products circulated in the United States during the Second World War.[80]

Folklorist Janet Langlois provides a revealing, detailed analysis of the harm that was done by the spread of 9/11 rumors to The Sheikh, a Mediterranean restaurant in a wealthy suburb of Detroit owned by a Lebanese immigrant (Langlois 2005). This story was covered widely in the local and national press. It begins with a widely recirculated email the day after the terrorist attacks by someone who reportedly saw employees at the restaurant cheering as the planes crashed into the Twin Towers and the Pentagon – much like the Dunkin Donuts and 7-11 rumors mentioned above. Copies of the email were also taped to shopping carts at a nearby supermarket (Zaslow 2002). The neighborhood in which the restaurant was located had a large

---

[78] Barbara Mikkelson, Dunkin Donuts 'Celebrating Employees' Rumor.

[79] David Mikkelson, Did Budweiser Pull Their Product from a Store Where Arabs Celebrated the 9/11 Attacks?, originally published as "This Bud's Not for You", originally published 18 October 2001, updated 10 September 2018, http://www.snopes.com/rumors/budweiser.asp

[80] Smith et al. (2010). Coca-Cola has long been the target of various rumors and urban legends, most frequently that it has bottled rodents in its products. See Cortada and Aspray (2019).

Jewish population, and Langlois was able to trace the circulation history of the email – something that is often difficult to do. The initial email was sent between two members of a local reformed Synagogue, a physician and his mother. The message circulated first within the membership of that synagogue and then across other reformed synagogues in the Detroit metro region before being circulated more widely to the public. (In fact, while there is a large Arab population in the Detroit metro area, there are actually few Arabs living in the neighborhoods near The Sheikh, which is more of an upscale restaurant than an inexpensive ethnic restaurant patronized by Arabs.)

The impact of the rumor on the restaurant was immediate: the number of patrons dropped by half overnight and was still reduced 4 years later. The restaurant received both letters and phone calls, expressing hope that the restaurant would go bankrupt. 18 of the 30 employees had to be laid off because business had slowed (Zaslow 2002). The owner, Noureddine Hachem, learned about the rumor from several regular customers who had received the email and called to inquire. Hachem's lawyer filed a defamation suit against the person believed to have written the original email, and the fact that a suit was filed intensified the boycott. Several members of the Arab-American community called for the defamation suit to be tried as a hate crime, and this further inflamed the controversy.

Hachem took various actions to counter the rumor. He compiled a list of names and telephone numbers from carryout receipts on September 11 and called these customers to see if any of them had seen any celebration of the terrorists occurring in the restaurant – none of them had. He also checked the restaurant's security videotapes for the entire day, and none of them showed employees celebrating. He decorated the restaurant with American flags and testimonials from customers. He advertised in the local paper, *Jewish News*. He asked local community leaders to speak up on his behalf, and the local Jewish Community Council, a number of synagogues, and the local Anti-Defamation League all showed him strong support, arguing that the Jewish community should follow The Ten Commandments and not bear false witness, and that these rumors were "just as evil as murder" (Zaslow 2002). Several Jewish leaders appeared on a live television broadcast from the restaurant, urging people to return. Several prominent members of the Jewish community made a conscious effort to eat regularly at the restaurant as a counter-boycott.

Hachem lost the suit and sold the restaurant in 2006. His family opened a new restaurant intentionally under a different name (Mediterranean Grill) at the Detroit Metro airport in late 2001 and parlayed that into concessions for 37 other restaurants at the Detroit and other major airports, including the franchises for Mrs. Fields cookies and Quiznos subs, as well as a sports bar, pizza joint, and Japanese restaurant (Henze 2014). Their company, Midfield Concession Enterprises, is today a $40 million business and currently employs 550 people.

3.7 Photoshops

'Photoshop' is the verb coined from the popular Adobe image-editing software to refer to digitally altered images – often humorous – used to provide a visual version of the urban legend known as "newslore". Internet folklore professor Russell Frank has made a detailed and insightful study of photoshops that appeared in the 2 months following 9/11.[81] More than half of these photoshops were centered on the theme of vengeance. Some were focused on annihilation of Osama bin Laden or Afghanistan, for example images of bin Laden in the cross-hairs of a rifle or Afghanistan shown in a cartoon map as having been bombed until it was nothing more than a parking lot. Others images were focused on the humiliation of bin Laden through references to sexual or scatalogical functions, such as a woman on top dominating bin Laden in bed in what was an intentionally emasculating image, or another of a roll of toilet paper with each paper square showing an image of bin Laden, variously with the text "Get Rid of your Shiite" or "Wipe Out Terrorism". These photoshops were similar to ones that had appeared earlier in connection with Saddam Hussein.

The other major category of 9/11 photoshops concerned victimization. The most famous example of these is Tourist Guy, a photoshopped image of what turned out to be a Hungarian tourist on the World Trade Center in which you can see a plane in the background that is about to crash into the building. For a short time, this image was believed to have been taken from a real photograph that had survived the destruction of the Twin Towers, but soon it was shown to be a hoax. This image became an Internet meme, and numerous variations of Tourist Guy quickly appeared. The Tourist Guy showed up in photoshops of the sinking of the Titanic, the Hindenberg crash, and the Kennedy assassination for example. In another photoshop in the same meme, it was Osama standing on the observation tower of the World Trade Center instead of Tourist Guy.[82] You can see our own version of this Internet meme in Fig. 3.1. It is an image that the everyday person can relate to – of being in the wrong place at the wrong time – and it helped ordinary Americans to come to terms with unspeakable horror in a way that the slick television and print news coverage did not help. One scholar said of these tourist images of 9/11:

---

[81] Frank (2004). For another thoughtful essay about the issues involved in archiving visual materials related to 9/11, see Hathaway (2005). Hathaway writes about the meaning of the photoshops discussed in this section and other visual responses to 9/11 that circulated online after the attacks: "Is it possible, I wondered, to come to any conclusions about the meaning of a piece of electronic lore when it arrives in your inbox with little, if any, introduction by the sender, and when the usual performance cues or other contextual information that might hint at the sender's motivation are absent? Do we receive e-lore in a contextual vacuum that the recipient must then "fill in" in order to make sense of the message? How do pre-existing contexts and identity markers affect the way we receive and interpret these items?" Hathaway also notes that while it may seem that textual jokes will have more limited geographic play because of linguistic constraints, somewhat surprisingly, she found with showing American versions of 9/11 photoshops to German students, that "visual jokes evoke a broader range of connotations and interpretations."

[82] See the Wikipedia article on Tourist Guy to understand the history and spread of this Internet meme.

Fig. 3.1 This photoshopped image, as many that follow this Internet meme do, begins with the original image of the Tourist Guy. In fact, the authors of this book were not on the observation platform of the World Trade Center when the planes crashed into the Twin Towers. Aspray was in his office two blocks away from the White House in Washington, DC but closed the office in case the White House was attacked, while Cortada was in his home office in Madison, WI. (Art created by Tara Knight and used with her permission)

> one hallmark of 9/11 narratives is an inability to process the events as part of everyday life; only through a cinematic filter can they be understood as "real." In this sense, 9/11 personal experience narratives capture the tourism dynamic at its most intense: the experience of passively witnessing the events was so powerful, so evocative of the "real," that it simulated the experience of being there in person (Hathaway 2005).

Hathaway also writes about the ironic dissonance involved in many of these 9/11 photoshops. With the exception of some people who believed the photograph of the Tourist Guy was a real, undoctored image, it is clear most of the time that the humor involved in these photoshops is knowing that what one is seeing is fake and understanding the "ironic dissonance between what one sees and what one knows to be "real" (Hathaway 2005).

In fact, the 9/11 attacks were not the first instance of victimization newslore. There had been a similar response to the Challenger disaster in 1986, when the Space Shuttle Challenger broke apart only seconds after takeoff, killing everyone on board in front of a live television audience of millions. The widespread availability of personal computers, photo editing software such as Photoshop, and the Internet has made it likely that photoshops will be a common part of the response to future disasters and their journalistic coverage.

3.8 Conclusions

The 9/11 terrorist attacks were followed almost immediately by numerous rumors and legends as people tried to make sense of the acts and to come to terms with the loss of a sense of safety and a way of American life. These rumors and legends provided a way for people to manage their fear by creating a narrative about a complex, dangerous, and incomprehensible situation, giving it a meaning that one can learn from or at least cope with. While the rumors were most widespread in the days and weeks after the attacks, they had a remarkable staying power. Ten years later, rumors were either still circulating or, more commonly, they had begun a new round of recirculation in either the same or a slightly modified form, displaying a resilience that had also characterized so many earlier rumors, from President Lincoln's assassination in 1865 to President John K. Kennedy's in 1963, all still circulating in the twenty-first century. There were hundreds of different rumors, falling into more than a dozen general categories. While the rumors were more frequently circulated in the regions in which the attacks occurred, especially in the New York metropolitan region, they were spread across the United States and beyond the borders of the nation. This was in part because of the ease of sharing stories online. But it was also because the fear that was pervasive in New York City also had resonance in other parts of the United States and the world that had previously been thought to be safe, to be distant from international terrorism.

The 9/11 attacks were a signal event for snopes. Together with the disputed Bush-Gore presidential election in 2002, the 9/11 attacks multiplied the number of rumors and legends circulating around the country. During these crises, snopes proved to be a reliable source and thus solidified its position as a leading fact-checker important to American public life. snopes was no longer simply a place for checking on urban legends and other curiosities. More importantly, snopes was part of an expanding fact-checking activity in American life that transcended the interests of a particular constituency, spreading broadly through American society at large. To illustrate that broadening of fact checking we need to discuss two further issues: the expanding role of scrutiny in other facets of American life and the increased intensified use of rumors and fact checking in political affairs. Those are the issues addressed in the next two chapters.

References

Aronson, Jay D. 2016. *Who Owns the Dead? The Science and Politics of Death at Ground Zero*. Cambridge, MA: Harvard University Press.

Behe, George. 1988. *Titanic: Psychic Forewarnings of a Tragedy*. New York: Antiquarian Press.

Bergen, Peter L. 2011. *The Longest War: The Enduring Conflict Between America and al-Qaeda*. New York: Free Press.

Bonaparte, Marie. 1947. *Myths of War*. London: Imago Publishing.

Brottman, Mikita. 2012. What's So Funny About 9/11. *Chronicle Review*, February 12, Chronicle of Higher Education. https://www.chronicle.com/article/Whats-So-Funny-About-9-11-/130708. Accessed 2 Dec 2017.

Burger, Peter. 2009. The Smiley Gang Panic: Ethnic Legends About Gang Rape in the Netherlands in the Wake of 9/11. *Western Folklore* 68 (2/3): 275–295.

Campion-Vincent, Veronique. 2003. The Enemy Within: From Evil Others to Evil Elites. *Ethnologia Europaea* 33 (2): 23–31.

Cortada, James W., and William Aspray. 2019. *Fake News Nation: The Long History of Lies and Misrepresentations in America*. Lanham: Rowman & Littlefield.

Degh, Linda. 2001. *Legend and Belief*. Bloomington/Indianapolis: Indiana University Press.

DeLillo, Don. 2008. *Falling Man*. New York: Scribner.

DiMarco, Damon. 2007. *Tower Stories: An Oral History of 9/11*. 2nd ed. Santa Monica: Santa Monica Press.

Dittmer, Jason. 2005. Captain America's Empire: Reflections on Identity, Popular Culture, and Post-9/11 Geopolitics. *Annals of the Association of American Geographers* 95 (3): 639.

Dutta-Bergman, Mohan J. 2006. Community Participation and Internet Use After September 11: Complementarity in Channel Consumption. *Journal of Computer-Mediated Communication* 11 (2): 469–484.

Dwyer, Jim, and Kevin Flynn. 2004. *102 Minutes: The Untold Story of the Fight to Survive Inside the Twin Towers*. New York: Times Books.

Ellefritz, Richard G. 2014. *Discourse Among the Truthers and Deniers of 9/11: Movement-Countermovement Dynamics and the Discursive Field of the 9/11 Truth Movement*. PhD Dissertation, Oklahoma State University. https://911inacademia.files.wordpress.com/2011/01/ellefritz-osu-phd-thesis.pdf. Accessed 18 July 2018.

Ellis, Bill. 2002. Making a Big Apple Crumble: The Role of Humor in Constructing a Global Response to Disaster. *New Directions in Folklore* 6. https://web.archive.org/web/20050303174535/http://www.temple.edu:80/isllc/newfolk/bigapple/bigapple1.html. Accessed 3 Aug 2018.

Foer, Jonathan Safran. 2006. *Extremely Loud and Incredibly Close*. Wilmington: Mariner Books.

Foot, Kirsten, Barbara Warnick, and Steven M. Schneider. 2005. Web-Based Memorializing After September 11: Toward a Conceptual Framework. *Journal of Computer-Mediated Communication* 11 (11): 72–96.

Frank, Russell. 2004. When the Going Gets Tough, the Tough Go Photoshopping: September 11 and the Newslore of Vengeance and Victimization. *New Media & Society* 6 (5): 633–658.

Gerould, G. H. (1908/2000). *The Grateful Dead: The History of a Folk Story*. Chicago: University of Illinois Press

Goldstein, Diane E. 2009. The Sounds of Silence: Foreknowledge, Miracles, Suppressed Narratives, and Terrorism – What Not Telling Might Tell Us. *Western Folklore* 68 (2,3 Spring, Summer): 235–255.

Gournelos, Ted, and Vivica S. Greene, eds. 2011. *A Decade of Dark Humor: How Comedy, Irony and Satire Shaped Post-9/11 America*. Jackson: University Press of Mississippi.

Gray, Richard. 2011. *After the Fall: American Literature Since 9/11*. Hoboken: Wiley.

Greenspan, Elizabeth L. 2003. Spontaneous Memorials, Museums, and Public History: Memorialization of September 11, 2001 at the Pentagon. *Public Historian* 25: 129–132.

Griffin, David Ray. 2004a. *The New Pearl Harbor: Disturbing Questions About the Bush Administration and 9/11*. Northampton: Interlink Pub Group.

———. 2004b. *The 9/11 Commission Report: Omissions and Distortions*. Ithaca: Olive Branch Press.

———. 2008. *The New Pearl Harbor Revisited: 9/11, the Cover-Up, and the Expose*. Northampton: Interlink Pub Group.

———. 2009. *The Mysterious Collapse of World Trade Center 7: Why the Final Official Report About 9/11 is Unscientific and False*. Ithaca: Olive Branch Press.

————. 2011. *9/11 Ten Years Later: When State Crimes Against Democracy Succeed*. Ithaca: Olive Branch Press.

Hagen, Kurtis. 2011. Conspiracy Theories and Stylized Facts. *Journal for Peace and Justice Studies* 21: 3–22.

Hathaway, Rosemary V. 2005. 'Life in the TV': The Visual Nature of 9/11 Lore and Its Impact on Vernacular Response. *Journal of Folklore Research* 42 (1): 33–56.

Hays, Charles A. 2011. The 9/11 Decade: Social Imaginary and Healing Virtual Community Fracture. *Global Media Journal – Canadian Edition* 4 (2): 79–94.

Heimbaugh, J. R. 2001. *Urban Legend Zeitgeist*. https://web.archive.org/web/20090312070141/http:/tafkac.org:80/ulz/. Accessed 3 Aug 2018.

Henze, Doug. 2014. Andrea Hachem: President and CEO, Midfield Concession Enterprises, Inc. *Crain's Detroit Business,* June 1; Updated March 16, 2017. http://www.crainsdetroit.com/article/20140601/FEATURE04/306019987/andrea-hachem. Accessed 28 June 2017.

Hofstadter, Richard. 1964. The Paranoid Style in American Politics. *Harper's Magazine*, 77–86.

Huddy, L., and S. Feldman. 2011. Americans Respond Politically to 9/11: Understanding the Impact of the Terrorist Attacks and their Aftermath. *American Psychologist* 66 (6): 455–467.

Ibrahim, Raymond. 2007. *The Al Qaeda Reader*. New York: Broadway Books.

Jackson, Kathy Merlock. 2005. Psychological First Aid: The Hallmark Company, Greeting Cards, and the Response to September 11. *The Journal of American Culture* 28 (1): 11–28.

Kean, T. 2011. *The 9/11 Commission Report: Final Report of the National Commission on Terrorist Attacks Upon the United States*. Washington, DC: US Government Printing Office.

Kuipers, Giselinde. 2005. 'Where Was King Kong When We Needed Him?': Public Discourse, Digital Disaster Jokes, and the Function of Laughter After 9/11. *Journal of American Culture* 28: 70–84.

Langlois, Janet L. 2005. "Celebrating Arabs": Tracing Legend and Rumor in Post-9/11 Detroit. *The Journal of American Folklore* 118 (468): 219–236.

Laqueur, Thomas. 2015. *The Work of the Dead: A Cultural History of Mortal Remains*. Princeton: Princeton University Press.

Lindahl, Carl. 2009. Faces in the Fire: Images of Terror in Oral Marchen and in the Wake of 9/11. *Western Folklore* 68 (2/3, Spring/Summer): 209–234.

Linenthal, Edward T., Jonathan Hyman, and Christiane Gruber. 2013. *The Landscapes of 9/11: A Photographer's Journey*. Austin: University of Texas Press.

Meder, Theo. 2009. They are Among Us and They Are Against Us: Contemporary Horror Stories About Muslims and Immigrants in the Netherlands. *Western Folklore* 68 (2/3, Spring/Summer): 257–274.

Melnick, Jeffrey. 2009. *9/11 Culture*. Hoboken: Wiley.

Morgan, Matthew J., ed. 2009. *The Impact of 9/11 on Politics and War*. New York: Palgrave Macmillan.

Ooze.com. 2006. *This 9/11 Joke Collection Will Save America! After 5 Years, You Can Laugh! It's Official*. http://www.ooze.com/articles/9-11-jokes.html. Accessed 12 Feb 2017.

Pedersen, Peter Ole. 2013. The Never-Ending Disaster: 9/11 Conspiracy Theory and the Integration of Activist Documentary on Video Websites. *Acta University Sapientiae, Film and Media Studies* 6: 49–64.

Peitz, William. 2013. *Captain America: The Epitome of American Values and Identity*. Senior Capstone Thesis, Arcadia University. https://scholarworks.arcadia.edu/cgi/viewcontent.cgi?referer=&httpsredir=1&article=1003&context=senior_theses. Accessed 5 Dec 2017.

Powell, Kimberly A. 2011. Framing Islam: An Analysis of U.S. Media Coverage of Terrorism Since 9/11. *Communication Studies* 62 (1): 90–112.

Pyszczynski, T., S. Solomon, and J. Greenberg. 2003. *In the Wake of 9/11: The Psychology of Terror*. Washington, DC: American Psychological Association.

Rosenblatt, Adam. 2015. *Digging for the Disappeared: Forensic Science After Atrocity*. Palo Alto: Stanford University Press.

Smith, Dennis. 2003. *Report from Ground Zero*. New York: Plume.

Smith, Trisha L., et al. 2010. The Grateful Terrorist: Folklore as Psychological Coping Mechanism. *Voices: The Journal of New York Folklore* 36 (Spring–Summer): 23–27.

Stempel, Carl, Thomas Hargrove, and Guido H. Stempel III. 2007. Media Use, Social Structure, and Belief in 9/11 Conspiracy Theories. *Journalism & Mass Communication Quarterly* 84 (2, Summer): 353–372.

Stevenson, Ian. 1960. A Review and Analysis of Paranormal Experiences Connected with the Sinking of the Titanic. *Journal of the American Society for Psychical Research* 54: 153–171.

———. 1965. Seven More Paranormal Experiences Associated with the Sinking of the Titanic. *Journal of the American Society for Psychical Research* 59: 211–225.

Waldman, Amy. 2011. *The Submission*. London: Picador.

Wong, Wing-Keung, Howard E. Thompson, and Kweehong Teh. 2011. Was There Abnormal Trading in the S&P 500 Index Options Prior to the September 11 Attacks? *Multinational Finance Journal* 15 (1/2): 1–46.

Wood, Michael J., and Karen M. Douglas. 2013. "What about building 7?" A Social Psychological Study of Online Discussion of 9/11 Conspiracy Theories. *Frontiers in Psychology* 4: 409, https://www.frontiersin.org/articles/10.3389/fpsyg.2013.00409/full. Accessed 18 July 2018.

Wright, Lawrence. 2006. *The Looming Tower: Al-Qaeda and the Road to 9/11*. New York: Alfred A. Knopf.

Yocom, Margaret. 2006. 'We'll Watch Out for Liza and the Kids': Spontaneous Memorials and Personal Response at the Pentagon, 2001. In *Spontaneous Shrines and the Public Memorialization of Death*, ed. Jack Santino. New York: Palgrave Macmillan.

Zaslow, Jeffrey. 2002. Arab's Restaurant is Nearly Ruined by Rumor of Celebration on Sept. 11. *Wall Street Journal*, 13 March, Reprinted Online by Bint Jbeil at http://www.bintjbeil.com/E/news/020313_sheik.html. Accessed 28 June 2017.

Zeitlin, Steve, and Ilana Harlow. 2001. 9/11: Commemorative Art, Ritual, and Story. *The Journal of New York Folklore* 27 (Fall–Winter). Available online at http://www.nyfolklore.org/pubs/voic27-3-4/dnstate.html. Accessed 2 Dec 2017.

Chapter 4
Debunking as Hobby, Entertainment, Scholarly Pursuit, and Public Service

The contents of [today's] legends – macabre happenings, accidents with household machinery, encounters with off-duty royalty, dreadful contaminations of foodstuffs, environment or bodily organs, theft, violence, threat, sexual embarrassments – seem to be novel in legendry. The setting of the time in the recent past and the reliance of the story on modern lifestyles seem further to separate these new forms from traditional legends. Hence we have come to call them 'contemporary legends'
-Bennett and Smith (1988)

4.1 Introduction

In the quarter century between 1990 and 2015, a number of seemingly unrelated events occurred. Hollywood released a raft of B-grade movies based on urban legends, such as *Candyman* and *I know What You Did Last Summer*. Both individuals and government agencies began to alert the public to scrutinize stories they found online to reduce the risk they would fall prey to online chair letters or computer viruses such as the Morris Worm. And folklorists and sociologists came together to create a new subdiscipline that studied the contemporary legends described in the epigraph to this chapter. These seemingly unrelated events in fact were not only connected to one another but also to the hobbyists described in Chap. 2 who actively participated in alt.folklore.urban. All of these activities involved taking a more critical, skeptical stance in an increasingly complex and dangerous world and in particular with an interest in contemporary legends as cautionary tales about how to survive in this dangerous environment.

The Internet has served as a place for both the uncritical spread and critical debunking of urban legends and other kinds of alternative facts and misinformation. This chapter surveys a number of the individuals and organizations that participated in these activities, covering primarily the quarter century from 1990 to 2015. These

© Springer Nature Switzerland AG 2019   77
W. Aspray, J. W. Cortada, *From Urban Legends to Political Fact-Checking*,
History of Computing, https://doi.org/10.1007/978-3-030-22952-8_4

activities began before there was an Internet as we know it today, on proprietary Internet services offered by America OnLine, Prodigy, CompuServe, the Source, and About.com. The chapter ends as the intense political use of alternative facts becomes pronounced, beginning with the 2008 U.S. presidential campaign and continuing to this day. The next chapter covers the history of political fact-checking.

These tensions between new ways to spread misinformation and new places to debunk these inaccuracies is an ongoing story. We believe that the examples we give provide the patterns of this tension. We have thus not tried to be comprehensive and follow the very latest trends, which themselves would have soon been succeeded by yet newer techniques. Thus, we have not covered patterns that were just emerging as we finished our manuscript, such as the discussion of urban legends and misinformation in the Ask Reddit section of reddit, the greentext stories on 4chan, or the numerous references to urban legends on Bustle. This chapter provides a context for the previous chapters about alt.folklore.urban and snopes, as well as the rumors and legends associated with the 9/11 terrorist attacks by showing some of the other related activities that were going on at the same time.

4.2   Pre-Internet Online Connections

In Chap. 2 we discussed the Usenet newsgroup alt.folklore.urban, which was created in 1991, the same year in which the World Wide Web was created, and the emergence of snopes.com from it 4 years later. These were the most dedicated places for urban legends to be scrutinized and possibly debunked, but they were not the place where most online urban legends were spread. Urban legends were spread primarily where ordinary people had their online connections. In the 1980s and into the 1990s, these venues were the proprietary online service providers such as The Source, Prodigy, CompuServe, and America Online – and even earlier in the 1960s and 1970s through non-commercial networks located largely in high schools, colleges, and universities.[1]

Before turning to the online spread of urban legends, we should note that other technologies were also used for the spread of urban legends – including the fax machine and even teletype machines – since at least the end of the Second World War. For example, the Lights Out legend concerned the distribution of allegedly official warnings to the public that gang members were driving around with their headlights off and would follow and murder any drivers who flashed their lights at them. Throughout the decade of the 1990s, this urban legend was spread by faxing official-looking warnings to friends, family, and co-workers. The rumor caused so much fear in Memphis that the police had to hold a press conference in 1993 and ask a local newspaper to publish a story indicating that there was no official publication on this subject and that there was no evidence of any such gang violence. A few

---

[1] See Rankin (2018).

months later, the Lights Out urban legend began to circulate in Chicago, through the medium of fax (Buck 2017).

In the 1980s and early 1990s, people gained online access most commonly by signing up with a specialized online service provider. These companies provided access to email (but only to write to other customers signed up with the same provider), chat, bulletin boards, and games and various kinds of information services. These services were at first available only in text format and were typically accessed through the use of a dial-up modem. The Source, launched in 1978 by founder Bill von Meister with *Reader's Digest* as its majority owner for most of its history, was the first online service provider focused exclusively on consumers. In the first half of the 1980s, The Source typically served between 20,000 and 60,000 members, with a peak of 80,000 members (Howitt 1984; *Online Timetable* n.d.). CompuServe, which was created in 1969 to provide batch processing and time-sharing services, used idle machine time in the evening during the 1980s to run an online service for consumers. CompuServe had the most technical clientele of all these providers and was owned by the tax preparation and business consulting firm H&R Block. CompuServe acquired The Source in 1989 (Webb 1989). By 1990, CompuServe had 600,000 users.

Another competitor, Prodigy, was founded in 1984 as a joint venture of CBS, IBM, and Sears Roebuck – with the companies investing over $600 million (Shapiro 1990). By 1990, Prodigy had 465,000 users, which in the 1990s grew to more than a million subscribers. It was a major innovator, including being one of the earliest companies to use graphical user interfaces online (although of a primitive variety) and to offer what today we would call "e-commerce". It received most of its revenue from online shopping and advertising rather than from memberships. Its email service, which had been designed originally to support its online shopping services, quickly evolved into a general email system. Prodigy also supported a heavily used message board system, and it was its email and message board services that enabled the spread of urban legends. Prodigy was the first of these early online service providers to offer full access (in 1994) to the World Wide Web. By 1999, Prodigy's original service had been eliminated in favor of modern Internet service (InformationWeek News Staff 1999). When that change took place, material from these early online providers was generally lost: "It had no where to go but away. That data was never on the Internet; it existed in a proprietary format on a proprietary network, far out of reach from the technological layman. It was then shuffled around, forgotten, and perhaps overwritten by a series of indifferent corporate overlords" (Edwards 2014).

America On Line (AOL) eventually became the largest of these online providers. In 1983, William von Meister started a company named Control Video Corporation, using the $1 million payout that he received when he left The Source. Control Video provided an online service for the Atari 2600 video game console. The company lasted only 2 years, but a new firm (Quantum Computer Services) was started from its remains in 1985, offering an online service (Quantum Link) to Commodore and Apple microcomputer users, and soon adding a way for IBM users also to access the service. In 1989, the company changed its name to America Online (AOL). It grew aggressively throughout the 1990s, representing half of the American households that had online connections. During this decade, the company formed alliances with various educational content providers, including the American Federation of Teachers, Discovery Network, Highlights for Kids, Library of Congress, National Education Association, National Geographic, Pearson, Princeton Review, Scholastic, Smithsonian Institution, and Stanley Kaplan, among others. Thus, AOL was both a conduit and a content provider. It was The Source, Prodigy, CompuServe, and especially America Online that provided the medium for the spread of urban legends.

One example of an urban myth spread through these online services was the Willie Lynch letter. This letter was supposed to have been the text of a speech given by a British slave owner from the West Indies to the James River colony in Virginia in 1712, teaching slave owners there how to understand and control their slaves.[2] Blogger and columnist Deborrah Cooper remembers her first encounter with this letter. She was more skeptical than many others who encountered this letter, as she explains:

> Sometime in 1991 I became a CompuServe client, a user of the very first email system for consumers....
> So here comes this Willie Lynch letter.
> I got it in my IN-box the first time shortly after I opened my CompuServe account, so this had to be late-1991 or so. I actually laughed when I read it, recognizing it as a hoax immediately. Majoring in communications, it was very easy to see that the phraseology and the terminology, cadence and sentence structure were decidedly 21st century.
> Yet, for some reason Black people like to think this letter is a real document, with historical references and everything though not one of them can say they did any research to verify its authenticity. This is when that magical word 'faith' comes into play.
> This letter can be nothing but a prank, a sick joke perpetuated by students from some unknown University or scientific research facility [because, as an earlier part of this blog states, these were the only places that had online access at the time]. They put it together and sent it out on CompuServe. That thing has been forwarded and re-forwarded *ad nauseum* ever since – almost 20 years later it[']s still floating around.
> Interestingly, every few years a new generation of Black Americans sees it for the first time and falls prey to the nonsense. However, there is no such thing as a Willie Lynch letter. There is no such person as Willie Lynch the slave trainer.[3]

In 1996, AOL created an Urban Legends website. Unlike alt.folklore.urban or snopes.com, it circulated urban legends organized into 15 categories, such as celebrities (e.g. Bill Gates as an obsessive coupon clipper) and death (stories about serial killer Jeffrey Dahmer). However, it made little to no effort to check the veracity of these urban legends. So, it was a site for dissemination, not scrutinizing and debunking (Cheng 1996). This Urban Legends website was part of AOL's The Hub, which was a partnership with New Line Television to reach college-aged males. The plan was for the most heavily visited of the web pages that AOL launched to be the inspiration for a television show created by New Line (Anon. 1996). AOL claimed it had 2.8 million people signed up for The Hub (Wired staff 1997). Urban Legends,

---

[2] The interested reader can find a copy in Anon. (2009).
[3] Cooper (2010). Also see Glanton (2006).

together with the rest of The Hub's content, was eliminated in 1998 when AOL moved away from creating its own content and began to aggregate the content of others. In some ways, this move to eliminate The Hub was not surprising inasmuch as college-aged males were not a well-represented demographic among AOL members (Festa 1998).

In early 2001 AOL merged with the media conglomerate Time Warner, the owner of major content providers such as *Time* magazine, Warner Brothers film studies, cable network HBO, Turner Broadcasting, and DC Comics. This pushed AOL even further toward being an entertainment provider rather than simply a conduit (online service provider). Although the Urban Legends website had disappeared from the AOL site, AOL continued to uncritically report urban legend narratives. For example, in 2016 one AOL article reported on the Slenderman horror myth as well as examples of the occult in the Denver airport, including the oversized horse sculpture at the entrance to the airport, which had fallen and killed the sculptor in his studio (true story) and was regarded by some as the death horse from The Book of Revelation in *The Bible* (Sowa 2016). Another AOL article that same year reported on an alleged portal to Hell in Blue Ash, Ohio, called Satan's Hollow (Stratford 2016). Both of these articles reported on these curiosities as an entertainment item rather than as serious investigative journalism.

In the 1990s and 2000s, brothers Bill and Rich Sones answered questions about urban legends on CompuServe.[4] Bill Sones coauthored with John McGervey a column on probability entitled *The Numbers Game*, syndicated by the *New York Times* and the *Los Angeles Times*.[5] With his brother Rich, who is a Ph.D. physicist, Bill Sones co-authored the book, *Can a Guy Get Pregnant?: Scientific Answers to Everyday (and not so Everyday) Questions* (Sones and Sones 2005). Together they also wrote a nationally syndicated column entitled Strange But True. Using science, the Sones brothers would answer strange questions that might be of interest to a general audience. The format was to answer questions supposedly from readers; however, the names of the supposed questioners give an indication that the questions were mostly created by the Sones brothers themselves. Table 4.1 provides a sample of the questions answered in the Strange But True column. Note the playfulness of the Sones brothers in who they attribute the questions to.

One early online site that paid close attention to urban legends was About.com. It was founded in 1996 in New York City as The Mining Company. Its business model was to establish webpages on numerous topics to provide expert advice and answers online. In the pre-Google era of the 1990s, About.com competed with Yahoo, AskJeeves, AltaVista, and Answers.com to answer people's questions online (Garun 2017). Each of the special topics on About.com had a human "guide" who was responsible for the content on the site. It was regarded as what we today call "infotainment", providing "a human-filtered network of special-interest links,

---

[4] strangetrue@compuserve.com

[5] bluebox 19, Interview with Bill Sones, Writer of Strange But True, *Teen Ink*, August 20, 2012. http://www.teenink.com/nonfiction/interviews/article/487844/Interview-with-Bill-Sones-Writer-of-Strange-but-True/ (accessed 12 December 2018).

Table 4.1 Examples of questions answered by Strange But True column[a]

Issue of *The Hook* in which the column appeared	Date	Question
101	Feb 7, 2002	In the climactic scene of the 1976 movie remake, King Kong falls from atop a New York skyscraper and lies dying in the street below, where Jessica Lange bids him a tearful goodbye. How is his demise wholly unrealistic from a "falling bodies" standpoint?
201	Jan 9, 2003	Is it really true that the only human-made thing big enough for astronauts to see from the Space Shuttle with the naked eye is the Great Wall of China? – J. Glenn
302	Jan 15, 2004	The ancient fear of being taken for dead and then buried alive has given way today to the nightmare of being kept alive indefinitely in a comatose state. But aren't there still circumstances where modern electrocardiography and electroencephalography erroneously fail to detect life? – E.A. Poe
440	Oct 6, 2005	Had my parents waited 5 min or maybe half an hour or longer before having sex that led to my conception, how might I differ from the person I am today? –M. Hyde
501	Jan 5, 2006	Animal engineering: echo-ranging by bats, dams built by beavers, parabolic reflectors of limpets, infrared heat-seeking sensors in some snakes, the hypodermic syringe of wasps and snakes and scorpions, the harpoon of cnidarians, and jet propulsion by squids. More could be cited, but not the wheel, that proverbial human invention. Why not? –M. Perkins
615	Apr 12, 2007	Oh, no, rats have invaded your dwelling! What might the varmints discover about your family that you don't know? –P.P.O. Hamelin
735	Aug 28, 2008	If angels really existed and could fly, from a structural standpoint what would they have to look like? –L.U. Cifer
802	Jan 15, 2009	If you ever get an opportunity to walk in space, be sure to wear your spacesuit. Because if you don't… what?–N. Armstrong
910	March 11, 2010	How does the Coast Guard find people lost at sea?– J.P. Jones
1009	March 3, 2011	*If the one-child-per-couple rule is strictly enforced in China, how long would it take before its population goes to zero? –M. T. Tung*
1203	Jan 17, 2013	If chimps played ball, would they throw right- or left-handed? –J. Goodall

[a]Source: *The Hook*, Charlottesville, VA (http://www.readthehook.com/columns/Strange%20But%20True)

feature stories, and chat events" (Kushner 1997). The site was renamed About.com in 1999 because it was thought the name change would be beneficial to advertisers and e-commerce partners (Fleming 1999). As of 2000, About.com had 700 topic sites, covering 50,000 subjects, with 4000 advertisers and 21 million visitors per month.[6] Primedia (publisher of *Seventeen*, *Modern Bride*, *New York*, and another

---

[6] "about.com", Wikipedia

200 magazines) acquired it in early 2001, followed by acquisition by the *New York Times* in 2005 and by the media conglomerate IAC (owner of Ask.com) in 2012 (CNN financial news staff 2000). About.com was renamed Dotdash in 2017 (Ingram 2017). The Urban Legends site moved to Thoughtco.com, which is a niche website of Dotdash devoted to education.

About.com had a topic site devoted to urban legends from 1997 to 2016. David Emery, a freelance writer who has worked for snopes since 2016, ran About.com. The About.com urban legends page has been continued on ThoughtCo. Emery's column on About.com received praise from various major newspapers and the Poynter Institute. His webpage today includes as subtopics: In the News, Classic & Historic Legends, Rumors & Hoaxes, Animal Folklore, and Scary Stories. See Table 4.2 for examples.

For the most part, these online means of communication died out when the commercial Internet became established in the 1990s. The major legacy of this early enthusiasm for urban legends lives on in the Strange But True columns of the Sones brothers and the archived and current pages of ThoughtCo.

Today, people complain about the Internet for its role in the widespread distribution of false information and the uncritical acceptance by a large part of the population of whatever they read online. However, as we saw in this section, the Internet is only the latest media to serve this function. Proprietary online providers,

Table 4.2 Examples of urban legends recently analyzed by ThoughtCo[a]

Category	Example	Date	Truth
In the News	Did HobbyLobby Really Close 500+ Stores Due to ObamaCare?	April 16, 2018	False, no stores closed.
Classic & Historic Legends	Was Baseball's 7th Inning Stretch Created When President William Howard Taft could no longer sit in a seat too small for him on Opening Day 1910?	May 25, 2017	False, originated with Manhattan College in 1882.
Rumors & Hoaxes	Did Little Mikey from the Life cereal commercials die from his stomach exploding when he ate a bag of Pop Rocks candy followed by a Coke?	March 6, 2017	The actor who played Mikey (John Gilchrist) is alive and well as an ad executive in the broadcast industry, but General Foods had to stop selling Pop Rocks because of the bad publicity.
Animal Folklore	Are there albino alligators in the sewers of New York?	May 5, 2017	It is too cold and polluted for them to live there. Occasionally, alligators are caught in New York City, mostly escaped or abandoned pets most likely.
Scary Stories	Crime ring in New Orleans drugs tourists, removes their kidneys, sells them on black market	March 18, 2017	Although this rumor spread wildly in New Orleans just before Mardi Gras in 1997, there is no evidence of any such activity.

[a]Source: ThoughtCo. Emery's About.com column

Usenet, bulletin boards, traditional media, and even fax played a similar function. Seemingly, whenever there was a new mass communication technology, people appropriated it for this use (among many others). Some of the dissemination was simply to share curiosities or done unthinkingly, but at times it was intended to help others not fall prey to a danger lurking in our complex world. We also saw from the popularity of the Sones brothers that many people have an interest both in learning about curiosities in our amazingly diverse world and being able to apply science and common sense to separate fact from fiction in it – and that it is fun doing so! Interactive formats enabled these activities; whether it was to participate in a social network, an interactive website, or a Usenet newsgroup (or even earlier in talk radio).

We have examined in Chaps. 2 and 3, as well as in this section, how technology was used to not only disseminate but also scrutinize urban legends. We next turn to other contemporaneous activities that were connected to urban legends, but not obviously so, by their engagement with the scrutiny of the complex, dangerous world in which we live. We first turn, in the next section, to the academic study of the cultural meaning of urban legends.

4.3   Scholarly Activity

Scholarly interest in urban legends, fake facts, misinformation, and related phenomena has primarily been the province of folklore scholars, although it has also been studied by sociologists, anthropologists, and library and information scholars. Folklore became a scholarly study in Germany, Britain, and Scandinavia and a little later in America in the nineteenth century. The academic formalization of this field of study came through the creation of national folklore societies in a number of Western countries in the second half of the nineteenth century. In the United States, the American Folklore Society was formed in the 1880s, at the same time other academic disciplines were organizing, such as historians with their American Historical Association. During the 1930s, as part of the New Deal Works Project Administration, out-of-work writers went into the field and collected regional oral folklore. The increased study of folklore in America was stimulated by the bicentennial passage of the American Folklore Preservation Act in 1976. American study of folklore focused on celebrating the different cultures of melting-pot America. Today, in the United States there are strong folklore studies programs at a number of universities, including California-Berkeley, George Mason, Indiana, Missouri, Ohio State, Utah State, and Western Kentucky, as well as subprograms within various anthropology and English departments.[7]

Our interest here is not in the general study of folklore but instead in the study of urban legends. In Chap. 2, we have already discussed the American scholar, Jan

---

[7] On the use of the World Wide Web for the study of urban legends, see Bacon (2011).

Brunvand, well known for his life-long study of urban legends.[8] The first known use of the phrase 'urban legend' appeared in the *New York Times* in 1925 in the sense of misinformation (Brunvand 2005). Between the two world wars, literary figures had pointed to the existence of urban legends in their writings, e.g. F. Scott Fitzgerald in 1925 in *The Great Gatsby* (Fitzgerald 1925) and Alexander Woollcott in 1934 in *When Rome Burns* (Woollcott 1934). Folklorists generally paid little attention to these urban legends, which they were not certain qualified as folktales and which were not easily analyzed according to their most commonly used theories and approaches. However, after the Second World War, urban legends began to attract the interest of some folklorists, sociologists, and other scholars. Here, folklorist and linguist John Widdowson explains the origins of urban legend research:

> Apocryphal anecdotes, urban legends, urban myths, contemporary legends – so, what's in a name?[9] The debate began in the 1970s, if not earlier, and continues to this day, as a hitherto apparently unrecognized genre of traditional narrative thrust its way into an unsuspecting world previously dominated by evolutionists, comparativists and others working on the types and motifs of international folktales. True, there were a select few who recognized the importance of legends – tales believed to be true or told as if true – but their research somehow failed to acquire the prestige accorded to the study of folktales proper. The arrival of an upstart in the form of a new and seemingly peripheral kind of legend was therefore

---

[8] Because of our focus on the United States, we do not tell here the story of the Centro per la Raccolta delle voci e leggende contemporanee and its journal: *Tutte Storie: Notiziario del Centro per la raccolta delle e leggende contemporanee.*

[9] Below is the definition of 'contemporary legend' given by Paul Smith, one of the leading practitioners of this field, However, as we discuss later in this section, Smith had a difficult time arriving at a definition that he believed was representative of this field of study and would pass muster with his colleagues.

> During the course of conversations with relatives, friends, and colleagues, we are often told stories about events which allegedly happened to a "friend of a friend" of theirs. Many of these stories are, of course, true. However, a proportion appears to be what are now recognized as urban legends.

Urban legends (alternatively described as contemporary, modern or belief, legends or myths) are short and highly mutable traditional narratives, or *digests* of narratives, which have no definitive texts, formulaic openings or closings, or artistically developed form, and so their traditional nature is not always immediately apparent. When communicated orally, they exist primarily as an informal conversational form, although they are also to be found embedded in other types of discourse (e.g. joke, memorate [an oral narrative from memory relating a personal experience], rumor, personal experience narrative, etc.), and in diverse settings–ranging from news reports to after-dinner speeches. They are also frequently disseminated through the mass media, novels and short stories, by E-mail, FAX and photocopier, and so have a wide international circulation.

Urban legends are primarily non-supernatural, secular narratives which are set in the real world. Told as if they happened recently, they focus on ordinary individuals in familiar places, and portray situations which are perceived as important by the narrators and listeners alike, and which they may have experienced, are currently experiencing, or could possibly experience. As such, they describe plausible, mundane, ordinary experiences and events, although often with an unusual twist. This *mundaneness* gives urban legends a unique quality which sets them apart from other forms of legend. Furthermore, urban legends emerge out of social contexts and interactions, and comment on culturally proscribed behavior. Such tales have been reported world-wide and their proliferation stands as a testament to their relevance in our society today…

unlikely to command instant acceptance, let alone respect, from an often hidebound and skeptical establishment in the world of academic folklore study. (Widdowson 2002)

Widdowson goes on to identify some of the doubts about urban legends expressed by these scholars and how their doubts challenged the creation of this new area of study:

And, the sceptic asks, what precisely is this "new" animal anyway? Isn't it the kind of story typically told at parties and other social gatherings by educated suburbanites? Doesn't it seem more at home in popular culture than in so-called "traditional culture", i.e. among "the folk"? Isn't it frequently mediated through print, photocopying, and more recently (horrors!) the internet? How can such a strange phenomenon be of interest to folklorists? Looking back over the extraordinary history of contemporary legend research it is easy to overlook the fact that questions such as these not only stood in the way of the birth of the subject but also closely reflected the deeply ingrained attitudes and mindset of many folklorists of the time. In these circumstances it is hardly surprising that the subject had such a difficult birth. (Widdowson 2002)

Widdowson then tells the creation story of urban legend research:

But maybe this was a blessing in disguise: it galvanized into action a handful of determined individuals who single-mindedly established the study of the genre and steered it purposefully forward over more than two decades, culminating in the creation of the International Society for Contemporary Legend Research. It all began when a small group of enthusiastic likeminded individuals met to participate in the first international conference, Perspectives on Contemporary Legend, in Sheffield in the summer of 1982. As the convener and prime mover, Paul Smith put it: "They discussed and argued … and constantly reassessed their own theories and research. The common bond was that … they had come with open minds to learn from one another. … we are now perhaps beginning to understand what the questions are." Are we? (Widdowson 2002)

For the remainder of this section we focus on one professional organization, the International Society for Contemporary Legend Research (ISCLR), founded in 1988. It is a small organization, which had only 82 members in 2006 and 55 in 2010, but it includes many of the leading scholars and most active amateurs in the field of urban legend scholarship (Henke 2010). The first meeting in the United States of scholars interested in contemporary legend was held in 1984 at the annual meeting of the American Folklore Society.[10] However, ISCLR had its origins in the annual seminar in Sheffield, England mentioned above by Widdowson, which had begun 2 years earlier.[11] The annual Sheffield conference evolved into the annual conferences of ISCLR, moving from place to place on both sides of the Atlantic from one year to the next.

---

[10] *FOAFTale* News No. 2 (June 1986), http://www.folklore.ee/FOAFtale/ftn01_10.pdf, accessed 20 June 2018.

[11] FOAFTale News No. 1 (September 1985), http://www.folklore.ee/FOAFtale/ftn01_10.pdf (accessed 20 June 2018). For the proceedings of the first Sheffield conference, see Smith (1984). Even earlier, there had been a one-time conference on urban legends held in 1969, organized by the folklorist Wayland Hand (See Hand 1971). It would take us too far afield to analyze the material covered at the early Sheffield conferences. Proceedings from the first four of these conferences are available in Smith (1984); Bennett et al. (1987); Bennett and Smith (1988, 1989).

What does ISCLR do? The Welcome page of its website indicates that the organization "encourages study of so-called 'modern' and 'urban' legends, and also of any legend that circulates actively." This page goes on to say that "members are especially concerned with ways in which legends merge with life: real-life analogs to legend plots, social crusades that use legends or legend-like horror stories, and search for evidence behind claims of alien abductions and mystery cats.[12]

ISCLR conducts its business in the way of many academic organizations. It holds an annual meeting; presents awards to both new and accomplished scholars; and publishes a journal (*Contemporary Legend*), collections of conference papers (*Perspectives on Contemporary Legend*), and a newsletter (*FOAFTale News*).[13] Not directly connected to the organization, but as a sign of the maturation of the field, the community has produced a bibliography of its growing literature (Bennett and Smith 1993).

Our review of the 84 newsletters published over the years by ISCLR gives us an indication of topics of ongoing interest to this community of scholars, e.g., aliens/ UFOs and Satanism were very common topics. Other fairly commonly discussed topics include academic legends about college students and professors, Elvis sightings, HIV/AIDS legends, crime legends, and sexual situations that broke cultural norms.

Note that ISCLR uses the phrase 'contemporary legend' rather than 'urban legend'. The latter phrase is much more commonly found in American public discourse, but it admits many meanings, including "not true"; whereas 'contemporary legend' is used primarily by scholars with a careful meaning that identifies certain kinds of form in the narrative.[14]

Members of ISCLR have pointed to the similarities of their object of study to the object of research by folklorists, despite the fact that many folklorists have been skeptical of the study of urban legends. In particular contemporary legend scholars point to the long history of contemporary legends and suggest that this domain of study is comparable to the kind of comparative cultural analysis that has become common in folklore studies.

> Contemporary ("urban") legends are one of the most pervasive forms of folklore in active circulation, but they are far from a modern phenomenon. The same processes of using

---

[12] Our chapter on 9/11 rumors and urban legends provides an excellent example of how a real terrorist attack generated legends that resonated with people's feelings and their efforts to cope with a complex, dangerous world. The Slenderman legend, which we do not cover except in passing in this book, provides causation in the opposite direction – of a legend leading to real-world actions, what the semiotician Eco (1976) and the folklorists Degh and Vazsonyi (1983) have called 'ostention' (https://contemporarylegend.org, accessed 20 June 2018); Eco (1976); Degh and Vazsonyi (1983). For another, earlier example of ostention in the contemporary legend literature, see the discussion of the 1989 Needling Whitey legend in which Black teenagers were reported to have stuck needles into randomly chosen White females on the Upper West Side of Manhattan (Ellis 1989). On ostension more generally, see Fine (1991).

[13] FOAF is an acronym created by the folklorist Dale (1978) for Friend of a Friend, as in a friend of a friend told me this story, which is a common provenance for urban legends (Dale 1978).

[14] For a more careful analysis, see Hobbs (2010).

narrative to communicate and negotiate anomalous experiences can be traced back thousands of years. Contemporary legends are contemporary to the teller and audience, not solely to the scholar. And what had been thought of as purely local narratives were found to exist in multiple manifestations throughout the world.[15]

On the other hand, critics of contemporary legend scholarship have criticized the lack of clarity in this domain of study, especially in its early years, say at the Sheffield conferences of the mid-1980s.[16] Looking back on these times in 2002, 14 years after the founding of ISCLR, Bill Ellis, one of the most active members of ISLCR, wrote:

> Critics of the "contemporary legend" scholarship of the 1980s have commented on how key figures like Gillian Bennett, W.F.H. Nicolaisen, Paul Smith, and myself found it difficult to define key terms and concepts. Response has ranged from dismissive (Alan Dundes, Heda Jason) to combative (Linda Dégh), while other scholars like Gary Alan Fine and Patricia Turner have dropped use of "legend" to rely on sociological research on 'rumor' to interpret similar material (Jason 1990). The inconclusive nature of this research, and the reluctance of scholars to build on it, could lead one to believe that these earlier meetings, however exhilarating for those participating, were quixotic in more ways than one....
>
> I suggest that the early Sheffield Seminars were right to raise these and similar issues again, this time using the term "legend," and that legend scholarship as a result can describe such cultural material more accurately than conventional sociological concepts of 'rumor' and 'truth claim'. Prof. M. Wilson, in summing up the discussion, commented:
> "One last point that I would mention is that it seems that myth has been equated with ignorance. The implication is that with knowledge myth would disappear. But I think that it is one of the characteristics of a lively myth that it will continue beside the facts, and that the demonstration of its falsity does not destroy it."
> Twenty years after the first Sheffield Seminar..., the 'lively' nature of our quarry continues to draw us on, in spite of its elusiveness – indeed, perhaps because of it. (Ellis 2002)

In 1991, the community debated the methods of contemporary legend scholarship. Bill Ellis described the criticism leveled by the Israeli folklorist Heda Jason in 1990:

> A brief but pointed critique by Heda Jason, chiding English-speaking scholars for having proposed 'contemporary legend' as a banner for research without having clearly defined what it does and does not cover. Legend scholars would be better advised, she suggests, to begin by examining, classifying, and ordering existing archival materials, such as the large contemporary legend file at the University of California. Once 'we know what we are dealing with,' she argues, folklorists can follow the traditional path of folkloristics to

---

[15] https://contemporarylegend.org, accessed 20 June 2018.

[16] In this chapter, we have been vague about the definition of 'urban legend' or 'contemporary legend'. Paul Smith, one of the major scholars in this field, attempted in 1995 to define the term and found it a difficult task. He circulated a part of his paper entitled "Defining the Contemporary Legend: Trials and Tribulations" in the ISCLR newsletter to get comment. His description of the definitional characteristics of the contemporary legend ran for many pages. He identified primary characteristics based on narrative status, form, structure, style, dissemination, narrators, context of narration, content, truth, belief, selection, meaning, and function. He concludes: "Having already made several attempts at such definitions, either for students in my classes, as part of my academic writings, or as background material for readers of my popular anthologies, I felt that this task would not take up too much of my time. How wrong I was" (Smith 1999). Also see Smith (2002).

planned fieldwork, documentation of typical performances, social context, channels of transmission, and other descriptive projects. 'Interpretations will have to wait until the data are assembled,' she concluded. (Ellis 1991b)

Ellis took exception to the moratorium on interpretation while this "spade-work" was being done. In contrast, he argued:

> Whatever else we may say about the defining characteristics of contemporary legend, it is <u>urgent</u>. It presses on us as something that has 'just happened' or is 'just about to happen', and it often requires a decision to take quick, decisive action to avoid danger. This action… can hardly be judged as 'illogical' or a 'delusion', as it frequently confronts a symbolic danger with a symbolic response….
>
>   The lesson: contemporary legends have their meaning and power as events unfold, not in the otherworldly calm of the archived and annotated text. We should see 'modern' legends as part of an age-old folk process, and many historical surveys of legend complexes wait to be done. Still, let's not let our work 'lose the name of action'. We should try to spot legends as they emerge and we be bold enough to comment on them while they – and we – are at the cutting edge.[17]

However, a decade after the Sheffield lectures, in the mid-1990s, Ellis somewhat conceded that the critics had a point to make and that the concept of contemporary legend had multiple meanings:

> Recent scholarship has suggested that the term 'contemporary legend' covers a plurality of narrative complexes. Some of these form stable networks of motifs and performance dynamics… In these complexes, historical research has helped clarify what is 'contemporary' about the latest harvest of texts, and what is inherent in the genre. (Ellis and Mays 1994)

In our examination of the ISCLR newsletter, we found evidence of questioning or implied criticism of contemporary legend scholarship on three other issues. At the 1993 ISCLR Seminar, held in Bloomington, Indiana, a panel of scholars discussed the boundaries of the genre of contemporary legend and a seminar was organized on the sociological uses of the concept of contemporary legend.[18] There was also questioning about the impact of mass media on contemporary legends: Does the mass media contaminate contemporary legend or is it simply a part of the dynamics of legend spread?[19] Finally, there have been questions raised about the place of theory in contemporary legend research. A newcomer to the 2012 annual meeting, held in Göttingen, Germany, noted what were for her three surprising impressions:

- surprised by the lack of theoretical papers;
- even more surprised by the general lack of scholarly/academic disagreement;
- general conformity to the idea that everything of theoretical significance has already happened some time ago (Kvartic 2013)

Jan Harold Brunvand, the best-known academic figure in this field, has argued (2004) that the urban legend is "vanishing":

---

[17] Ellis (1991b). Also see Ellis (1991a).

[18] *FOAFtale News*, No. 30 (June 1993), http://www.folklore.ee/FOAFtale/ftn30.pdf accessed 21 June 2018.

[19] On the relationship between mass media and contemporary legends, see Degh (1994).

In my paper "The Vanishing 'Urban Legend'" (ISCLR 2001 conference) I suggested that "the 'urban legend' has much less vitality as an oral-narrative genre than in its glory days from the 1960s through the 1980s" and that examples "have mostly migrated from folklore into popular culture where they are stereotyped, standardized, exploited, commodified, and re-packaged [with] the most common medium for their circulation [being] the Internet." Not that there's anything wrong with that; and not that urban legend scholars lack data still to collect and study.

In my paper "'Urban Legend' as a Household Phrase" (ISCLR 2003 conference) I presented an informal review of how dictionaries and the mass media have adopted the term and modified its meaning. Examples ranged from items published in *The New Yorker* to *Weekly World News* and included such oddities as a question on the TV game show *Jeopardy*, several Hollywood films, and references in the press to a football coach whose first name is "Urban". Not that there's anything surprising about that; all anyone needs to do is "Google" the term "urban legend" on the Internet to discover how widespread its use has become.

This paper continues my survey of how the genre and the term "urban legend" flourish in the mass media and on the Internet, even while continuing in a seeming decline as an oral-narrative tradition. Examples, again, range widely to include comic strips, a cross-word puzzle, a board game, various television series, a neoconservative website, a murder mystery written by a Japanese author, and more. (Brunvand 2004)

We close this section with an account from the Australian folklorist Mark Moravec of the 12th annual ISCLR conference, held in Paris in 1994, which gives a snapshot of the mature yet still evolving scholarship in contemporary legend research. Moravec first points to the influence of Brunvand, the internationalism, and the multitude of academic disciplines represented at the conference:

> Here was a multi-disciplinary gathering of folklorists, psychologists, sociologists, historians, journalists and authors, each in our own way searching for the truth behind the legend process. Being a truly international conference, the informal discussions were often a multi-lingual affair in French, German, Italian and many different accented forms of English. The impact of Jan Brunvand's popularisation of legend studies was clearly evident in the wonderful array of multi-lingual books on display, most authored by the conference participants, and all charting the spread of contemporary legends around the world. (Moravec 1995)

Moravec then identifies some of the major themes from the conference:

- We can and should look at the possible historical derivations of contemporary legends, recirculating age-old motifs anew.
- As legend scholars, we should welcome the multiplicity of methodologies. We need both the detailed case study and wider sociological comparisons.
- Unexpectedly, many papers dealt with personal experience stories, as well as the more familiar friend-of-a-friend accounts.
- The content, distribution, process and popular appeal of the story are more important than whether or not the incident ever happened. However, the legend scholar has a responsibility to publicly identify what is true or false if this becomes known.
- Issues of definition frequently arose. One view, though not unanimous, was that we do not need to distinguish between legends and memorates, as the processes are more important than the types of claims. Gary Butler's "legend complex" was seen as a term which usefully sidesteps the never-ending definitional debates.

Finally, Moravec reflects on the meaning of contemporary legend:

> [Here is] my own opinion on the central meanings and functions of many contemporary legends. I would include here fears of the darker side of life, of personal violence, of untrustworthy technology, of food contaminations, and of an unpredictable and uncontrollable outside world that threatens to undermine the individual's control of his or her own life. Yet there is also the opposite pole of hilarity, entertainment, one-upmanship, surprise and good luck. At a deeper level, there is the age-old battle of good versus evil, and the possibility of personal salvation despite the obstacles of everyday life. From one perspective, contemporary legends could be likened to modern-day parables that both warn and educate. They are legends that resonate with historical echoes, ultimately deriving their power from age-old motifs that are still relevant in contemporary society. (Moravec 1995)

While the study of urban legends by the members of AFU that we discussed in Chap. 2 may seem like no more than a colorful hobby, the creation of a new academic discipline to study these urban legends suggests their importance. The scrutiny in this case is not about determining the truth and falsity of these legends but instead about identifying their cultural meaning. The fact that we live in a world of terrorism, computer viruses, and highly contested politics where politicians do not play by the rules of truth, and where new information and misinformation reaches the individual at "Internet speed" from all across the world is a scary proposition. This fear is expressed by these urban legends, and the sociologists and folklorists of contemporary legend theory have developed ways to uncover and analyze the cultural meaning of these legends.

4.4   Public Service – Computer Viruses, Chain Letters, and Internet Hoaxes

In this section we profile organizations that provide urban legend or hoax debunking as a public service. It is sometimes difficult to draw the line between organizations that carry out this mission as a hobby and those that do it as an official public service. Clearly, Hoaxbusters, as a website sponsored by the US federal government, is an official public service. Sophos, a leading international IT security product and service firm, offers a Hoaxes page that provides a list of the latest hoaxes as well as a list of older hoaxes organized alphabetically.[20] Symantec, the U.S. IT security leader, also maintains a Hoaxes Dashboard, which provides information on more than 100 hoaxes – some of which are more than a decade old, others current.[21] In the case of both Sophos and Symantec, these web pages provide a public service but, of course, also create good will for their company. AFU and snopes (before it became so commercialized) fall on the hobby side of this divide.

---

[20] https://www.sophos.com/en-us/threat-center/threat-analyses/hoaxes.aspx,   accessed   25   June 2018.

[21] https://www.symantec.com/security-center/risks/hoaxes

Even more clearly on the public service side, although also as an exercise in self-publicity, is Scambusters. Audri and Jim Lanford earned their living by being Internet marketing consultants and digital publishers, and by running their own websites for nutritional supplements and paperless solutions.[22] In 1994, they established the website Scambusters.org, which their website claims has helped over 11 million people avoid internet scams, identity theft, and hoaxes. They have published a series of articles online about various kinds of scams, such as those involving credit card fraud, fake loans, ransomware, being cheated when selling valuables online, donation and Medicare scams, fake overdue bill threats, and scams involving student debt relief. They also maintained web pages with tips about avoiding identity theft, and a Hoaxes Resource Center that listed approximately 50 common online hoaxes. In a tragic turn of events, they were killed in 2018, when heavy rains caused a landslide crushing their house near Boone, North Carolina.[23]

Another example of public service is the Urban Legend Combat Kit, which is the creation of Patrick Crispin, an educational technology specialist who manages the University of Southern California's Center for Scholarly Technology. The combat kit provides advice on installing an online tool on your personal computer, called a bookmarklet, that enables one to type in keywords when reading websites or doing email, and the tool searches on these keywords on the snopes website to see whether the stories they are encountering are factual.[24]

While our main interest is in organizations operating in the United States, there are organizations doing similar work in other countries. One example is Hoax-Slayer, which has been run by Brett Christensen in Australia since 2003. Another is ThatsFake.com, active since 2009, and its sister site, ThatsNonsense.com – both run by Craig Charles Haley in Somerset, England to debunk hoaxes, rumors, and fake news online.[25] Another is the Hungarian IT security products and services company VirusBuster, formed by Juliana Bozso and others in 1997. VirusBuster's website includes information about hoaxes and their spread, including information about how to recognize a hoax online as well as a list of some of the more common hoaxes such as It Takes Guts to Say Jesus.[26]

We next turn to a government organization (CIAC) that was a leader in protecting American organizations from online dangers such as the Morris worm. In late 1988, the Internet worm (a.k.a. Morris worm) appeared, creating havoc on government, business, and individual websites.[27] The worm was created and placed online by a

---

[22] See their website paperitis.com.

[23] https://www.wataugademocrat.com/news/victims-of-pine-ridge-house-collapse-identified/article_b86464f0-bfdb-58a4-88d1-2075033ee7eb.html

[24] http://netsquirrel.com/combatkit/ accessed 25 June 2018.

[25] See http://www.thatsnonsense.com/about/ (accessed 11 July 2018) for more details.

[26] See for example, VirusBuster's Hoax page for 29 May 2008 at https://web.archive.org/web/20080529001506/http://www.virusbuster.hu:80/en/viruslab/hoaxes/index

[27] For general information about the worm, see Eisenberg et al. (1989); Eugene H. Spafford, The Internet Worm Program, Purdue Technical Report CSD-TR-823, November 29, 1988 (revised December 8, 1988) https://spaf.cerias.purdue.edu/tech-reps/823.pdf (accessed 11 December 2018).

Cornell University graduate student, Robert Tappan Morris, who became the first person convicted of a felony under the Computer Fraud and Abuse Act. The country was not well prepared for this attack, and individual organizations struggled to find their own solution. The U.S. Government Accountability Office estimated that the damage may have been as high as ten million dollars.

A few weeks later, the Software Engineering Institute at Carnegie Mellon University formed the CERT Coordination Center to be a place where information on these computer security issues could be collected and disseminated. Soon after that, in 1989, the US Department of Energy (DOE) together with the US Air Force created the Computer Incident Advisory Capability (CIAC) team to carry out a similar function as CERT for its clients, in particular advising on computer incidents, providing virus and hoax alerts, and generally advising on computer security vulnerabilities (Wikipedia n.d.). Similar incidence response teams were formed by other organizations, but the reaction later that same year to the Wank worm made the computer security community recognize that greater coordination was necessary. CIAC was a founding member of a new coordinating organization formed in 1990, named GFIRST (Government Forum of Incident Response and Security Teams), and later an active member in the international organization FIRST. CIAC was renamed DOE-CIRC in 2008.[28]

Bill Orvis, working as an employee at Lawrence Livermore National Laboratory, was hired by the DOE to monitor hackers and malicious code in 1994 (Weisbaum 2007). When CIAC received numerous questions about questionable emails, it decided in February 1995 to set up a website to monitor and report on these kinds of incidents rather than answer each individual inquiry separately. By the time CIAC revised its website, in November 2000, the site had already received visits from almost 6 million people (CIAC and the U.S. Department of Energy 2000). The pace of incidents increased greatly after the 9/11 terrorist attacks in 2001 (Moyer 2001). Orvis also served as an expert for the media and appears in a number of news articles, giving advice on such topics as fake emails and the security concerns about online chain letters (Moyer 2001; Fraser 2003).

On the CIAC website, hoaxes were organized into eight categories, as described below in Table 4.4. These categories focused on hoaxes that one would encounter online. The first hoax category mentions the Good Times hoax, which we will look at in detail before looking at the CIAC hoax categories more generally. Similar hoaxes to Good Times had probably been circulating since the 1980s. CIAC alerted the community to The Good Times hoax in December 1994.[29] The original hoax had received wide circulation in November and December 1994 with the following text:

Here is some important information. Beware of a file called Goodtimes.
    Happy Chanukah everyone, and be careful out there. There is a virus on America Online being sent by E-Mail. If you get anything called "Good Times", DON'T read it or download

---

[28] https://www.first.org/about/history, accessed 31 May 2018; Wikipedia CIAC; https://www. us-cert.gov/about-us
[29] FN CIAC Notes 94-04c.

it. It is a virus that will erase your hard drive. Forward this to all your friends. It may help them a lot.[30]

Soon after the first CIAC posting, a new version of the Good Times hoax appeared, as follows:

> The FCC released a warning last Wednesday concerning a matter of major importance to any regular user of the InterNet. Apparently, a new computer virus has been engineered by a user of America Online that is unparalleled in its destructive capability. Other, more well-known viruses such as Stoned, Airwolf, and Michaelangelo pale in comparison to the prospects of this newest creation by a warped mentality.
>
> What makes this virus so terrifying, said the FCC, is the fact that no program needs to be exchanged for a new computer to be infected. It can be spread through the existing e-mail systems of the InterNet. Once a computer is infected, one of several things can happen. If the computer contains a hard drive, that will most likely be destroyed. If the program is not stopped, the computer's processor will be placed in an nth-complexity infinite binary loop – which can severely damage the processor if left running that way too long. Unfortunately, most novice computer users will not realize what is happening until it is far too late.[31]

So, CIAC issued a second notice – in April 1995.[32] It pointed out that the Federal Communications Commission (FCC) had never issued a warning, and that the FCC never would have done so because this kind of work was not in their purview. Note that this new version attributed the hoax to an America Online user. It was typical for savvy online users to cast aspersions on the supposedly unsophisticated users of America Online. This kind of frustration with uncritical Internet users was captured in a spoof of the Goodtimes Hoax, allegedly written by the American fantasy writer Patrick Rothfuss, given in Table 4.3.

Some of the other Malicious Code Warnings revealed on the CIAC website (see Table 4.4) in the late 1990s or the year 2000 included ones about real Trojan attacks (PKZ300), virus hoaxes (Death Ray Virus; A.I.D.S. Virus, but note the fact that it appears with this name at the heightened concern about the human AIDS virus; Irina Virus, which was a gone-wrong publicity stunt by a publishing company; Blue Mountain Cards, which was a version of mercantile attack), harmless Trojan (ghost. exe), Trojan hoax (PENPAL GREETINGS!; Bug's Life ScreenSaver), and a chain letter hoax (AOL V4.0 Cookie, but note again the attack on AOL).

We will not analyze all of the different categories of hoaxes followed by CIAC. However, a few words about those hoaxes that CIAC labels as urban myths are in order to see how they compare with those discussed on AFU and snopes. Of the 26 urban myths listed on the CIAC Urban Myths and Legends page of 13 December 2000,[33] three are common ones that we saw on both AFU and snopes: the Klingerman

---

[30] All of the material, including the direct quotations about Goodtimes given below are from the 2 December 2000 CIAC webpage entitled "Malicious Code Warnings," https://web.archive.org/web/20001202032500/http://hoaxbusters.ciac.org:80/HBMalCode.shtml#goodtimes, accessed 11 June 2018.

[31] CIAC, "Malicious Code Warnings".

[32] CIAC Notes 95–09.

[33] https://web.archive.org/web/20001213145400/http://hoaxbusters.ciac.org:80/HBUrbanMyths.shtml#klingerman, accessed 11 June 2018.

Table 4.3 Goodtimes hoax spoof (Christensen 2016)

Goodtimes will re-write your hard drive. Not only that, but it will scramble any disks that are even close to your computer. It will recalibrate your refrigerator's coolness setting so all your ice cream goes melty. It will demagnetize the strips on all your credit cards, screw up the tracking on your television and use subspace field harmonics to scratch any CD's you try to play.

It will give your ex-girlfriend your new phone number. It will mix Kool-aid into your fishtank. It will drink all your beer and leave its socks out on the coffee table when there's company coming over. It will put a dead kitten in the back pocket of your good suit pants and hide your car keys when you are late for work.

Goodtimes will make you fall in love with a penguin. It will give you nightmares about circus midgets. It will pour sugar in your gas tank and shave off both your eyebrows while dating your girlfriend behind your back and billing the dinner and hotel room to your Discover card.

It will seduce your grandmother. It does not matter if she is dead, such is the power of Goodtimes, it reaches out beyond the grave to sully those things we hold most dear.

It moves your car randomly around parking lots so you can't find it. It will kick your dog. It will leave libidinous messages on your boss's voice mail in your voice! It is insidious and subtle. It is dangerous and terrifying to behold. It is also a rather interesting shade of mauve.

Goodtimes will give you Dutch Elm disease. It will leave the toilet seat up. It will make a batch of Methanphedime in your bathtub and then leave bacon cooking on the stove while it goes out to chase gradeschoolers with your new snowblower.

Listen to me. Goodtimes does not exist.

It cannot do anything to you. But I can. I am sending this message to everyone in the world. Tell your friends, tell your family. If anyone else sends me another E-mail about this fake Goodtimes Virus, I will turn hating them into a religion. I will do things to them that would make a horsehead in your bed look like Easter Sunday brunch.

Virus, which is a physical (not a computer virus) that one is exposed to by opening a package labeled "A gift for you from the Klingerman Foundation"; The Bad Guy in the Backseat urban legend, about gang initiation to attack women pumping gas at night; and the harvesting of kidneys of business travelers. Others concern adulterated foods (Flesh Eating Bananas, allegedly a warning from the Center for Disease Control; Kentucky Fried Hoax, what is in that "chicken" now that the company has renamed itself KFC?; and roaches in Taco Bell products), unsafe everyday products (anti-perspirant and shampoo cause cancer; sunscreen causes blindness), online urban legends (Internet Access Charges; E-Mail Tax Hoax), danger in everyday life from needles (on theater seats, in playgrounds, on gas pump handles) and poison (on pay phone keys and bank teller envelopes), and anti-religious activities (Federal Communications Commission ban on religious broadcasting; Proctor and Gamble promotes Satanism).

The CIAC website was closed in early 2008, after 13 years of operation, when the US Department of Energy decided it no longer wished to support it.[34]

There were other sites that pursued fake stories related to the spread of viruses or the harvesting of email addresses. The final example we consider is VMyths. This

---

[34] https://web.archive.org/web/20080531135314/http:/hoaxbusters.ciac.org:80/

Table 4.4 CIAC hoax categories[a]

Hoax category	Description
Malicious code (virus and trojan) warnings	Warnings about Trojans, viruses, and other malicious code that has no basis in fact. The Good Times and other similar warnings are here.
Urban myths	Warnings and stories about bad things happening to people and animals that never really happened. These are the poodle in the microwave and needles in movie theater seats variety.
Give-aways	Stories about give-aways by large companies. If you only send this on, some big company will send you a lot of money, clothes, a free vacation, etc., etc. Expect to wait a long time for any of these to pay off.
Inconsequential warnings	Out of date warnings and warnings about real things that are not really much of a problem.
Sympathy letters and requests to help someone	Requests for help or sympathy for someone who has had a problem or accident.
Traditional chain letters	Traditional chain letters that threaten bad luck if you do not send them on or that request you to send money to the top *n* people on the list before sending it on.
Threat chains	Mail that threatens to hurt you, your computer, or someone else if you do not pass on the message.
Scam chains	Mail messages that appear to be from a legitimate company but that are scams and cons.

[a]Source: https://web.archive.org/web/20001213062800/http://hoaxbusters.ciac.org:80/HBHoax Categories.html (accessed 11 June 2018) Reformatted but quoted verbatim

website has its origins in a "Computer Virus Myths treatise", which led to the creation of a website named Computer Virus Myths in 1995. The website was renamed VMyths.com in 2000. The site was closed in 2003 because of low advertising revenue to support the operation and the deployment of the founder, Rob Rosenberger, to the Persian Gulf by the US Air Force (MonsieurEvil 2003). The stated purpose of the website was to "dispel computer security myths, misconceptions, urban legends, and hoaxes…[and] secondarily to improve the reader's knowledge of lesser-known, technically oriented, and/or controversial issues in computer security hysteria." As part of the website philosophy, "Comedy is encouraged because it is the single most effective weapon against hysteria" (VMyths 2008). Or as one of the fans of the site described it: "A real one-stop shop for debunking nonsense spewed by hyperactive teens and TV news magazines…" (MonsieurEvil 2003). The site included both an archive of computer virus hoaxes and regular commentary by Rosenberger and others on hype in the computer security industry. Rosenberger was a regular speaker at conferences organized by CERT and CIAC.

Here is an example of Rosenberger's sarcasm, which he applied when defining the *addictive update model*.

> A subscription-based business model. The seller provides a product that requires regular maintenance and the buyer pays for a periodic subscription to receive said regular maintenance.
> The addiction comes from the feeling of euphoria the buyer gets after receiving an update. Most used by anti-virus companies, an update offers feelings of contentment and

security in the belief that the buyer's computer is safe from viruses. The high eventually wears off, however, as new variants of viruses appear, and the buyer must obtain another update. This exploits another aspect of addiction: *The buyer is not in control.*

This business model is unique to the information technology industry. While real world similarities exist, such as extended warranty, warranty subscriptions eventually expire, not to be extended. An addictive update model subscription is theoretically perpetual, provided the buyer continues to subscribe. (gordonf 2004)

The editors of the VMyths website were Rob Rosenberger and George C. Smith. Smith was the editor of *Crypt Newsletter* and author of the book *The Virus Creation Labs* about the early days of viruses (Smith 1994). He is a PhD chemist who studied flesh-eating bacteria and held a journalism fellowship from the Knight Foundation to study the military-industrial complex. He is also the creator of the website Dick Destiny, which has the tag line: "Expert Rock and Roll Ratings and Expert National Security Affairs Stories by George Smith, not necessarily in Any Order or Absolutely Guaranteed."[35] Rosenberger wrote one of the columns for the website.[36] He was an early anti-virus expert in the 1980s and had extensive computer experience with both industry and the federal government, having worked in the Air Force and with the CIA. He uses sarcasm and satire to critique both computer security professionals and the news media in cases where he believes they have over-hyped a problem or created or distributed misinformation about viruses. Not surprisingly, he is not particularly well liked in the computer security industry; but he does serve as a conscience for it and makes professionals think twice about what they communicate to the public about viruses; people admire his professionalism and willingness to admit his own mistakes (Delio 2001). As Rosenberger said about two common virus hoaxes, "It wasn't [end users] who passed the message. [It was] people working with Unix and NT boxes who know a little about viruses…There are so many new people on the 'net who read about evil geniuses who spread viruses in a temper tantrum and they believe it. So when they see an alert, they think they have to forward it."[37]

The organizational scheme for Vmyths is interesting. "Myths" are listed by sector, by type of hysteria, and by victims of hysteria. Sectors include customers, government and military, reporters and bloggers (with a special call-out for Wikipedia),

---

[35] For example, see the hypothetical remake of the Strangelove story in a June 9, 2008 posting entitled "Chinese Cybermen Said to be Impurifying our Precious Bodily Fluids: Cyber-Arclight Strikes Eyed as Remedy" (Smith 2008).

[36] The other columnists for VMyths were Robert Vibert, Lewis Koch, and someone who wrote under the pseudonym Vea Culpa. Vibert once led the anti-virus team at Sensible Security, the leading Canadian anti-virus organization, and has also worked as a computer security writer and consultant in Europe and Canada and more recently as an author and consultant on Awareness Expression Resolution, a system designed to promote mental and physical healing; he also served as the administrator for the Anti-Virus Information Exchange Network. Lewis Koch is a prize-winning investigative journalist who directed the Urban Journalism Fellowship Program at the University of Chicago in the 1970s and has taught at Columbia University and other colleges. He has covered cyber-issues as a special correspondent for CyberWire Dispatch (Moser 2007; Gosztola 2007).

[37] Rosenberger quoted in Goff (1998).

security vendors (with a special call out for their press releases), and think tanks. Types of hysteria include "solar calculator" math (i.e. myths involving numerical predictions and claims), chain-letter emails and tweet, charlatans and false prophets, cyber-war and cyber-terror, media circus, and Supervisory Control and Data Acquisition (SCADA).

VMyths was an interesting site. Not only did it provide a ready reference to misinformation online, but it also served as a reflective critic of the debunkers and journalists, indicating times where they went overboard in their calls for concern.

The scrutiny provided by CIAC and VMyths did not focus on the truth or falsity of claims. Instead, the point was to get people to scrutinize online activities more closely so that they did not have bad things happen to them such as accidentally uploading a computer virus to their computer or to their organization's computer system. In these cases, the truth or falsity of claims made on a website or in a received email were merely a tool to enable one to protect oneself online.

4.5   Entertainment

Finally, we briefly consider urban legends as the basis for entertainment as employed in two media, television and film.[38] In the case of television, we are most interested in television series in which a narrative is told and the audience is asked to determine whether it is true or false – with revelation at the end of the segment. This was the basic format for all five shows that we examined. An example was the documentary-style TV series *Urban Legend*, which featured three stories on each 30-min episode. Actors were employed in the case of fake stories, while the actual participants were interviewed in the true stories. The series aired in the United States on the Biography Channel in 2007 and on the SyFy channel in 2011. It can be found occasionally today in reruns.[39]

Mythbusters was a television show first pitched to the Discovery Channel by an Australian producer, Peter Rees of Beyond Productions. Discovery Channel originally turned down his proposal because it was already planning a show with similar content. The concept was tweaked so that, instead of simply telling the stories, a new element was added: to use scientific methods to test important elements of the rumor, legend, or myth. This reconceived format was picked up in the United States by the Discovery Channel, turned out to be popular, and ran for

---

[38] We have already discussed in Chap. 2 some of the books and newspaper columns about urban legends, especially the multiple books of Jan Brunvand. We have also discussed in that chapter and earlier in this chapter some of the other syndicated columns that appeared in newspapers and magazines about debunking stories, such as those by the Sones brothers. Another example is the Myths and Legends podcast since 2015 by Jason Weiser, which "re-tell[s] stories from myths, legends, and folklore from all around the world." (https://www.mythpodcast.com/where-to-start/) There are dozens of other urban legend podcasts. Find a list at https://player.fm/podcasts/Urban-Legends. On comic book urban legends, see https://www.google.com/search?client=safari&rls=en &q=comic+book+urban+legends&ie=UTF-8&oe=UTF-8

[39] "Urban Legends, TV Series", *Wikipedia*, accessed 26 June 2018.

fourteen seasons (2003–2017). The Science Network, a partner channel to the Discovery Channel, ran the show for one additional year under the name *MythBusters: The Search*.[40]

Mostly True Stories?: Urban Legends Revealed was a documentary-style television show shown on The Learning Channel for four seasons (2002–2004). In addition to reenactments of the legends and the revelation at the end, before the commercial break the audience was asked a true-false question such as whether George Washington cut down a cherry tree, with the answer given after the commercial. The show did answer one question incorrectly during the 2003 season, about whether Blackbeard used the nursery rhyme *Sing a Song of Sixpence* as a code to recruit pirates. snopes revealed afterwards that it had circulated this false story.[41]

Beyond Belief: Fact or Fiction was a television series produced by Dick Clark Productions and run on the Fox Network from 1997 to 2002. The show was unlike the others mentioned above in that it more consistently focused on stories of psychics or the supernatural. Many of the stories on the show were retellings of classic urban legends. Four or five stories were shown in each episode, typically with two or three being true. The show is available today on Amazon Prime Video.[42]

Finally, *Unsolved Mysteries* was a television program shown from 1987 to 2002 and again from 2007 to 2010 sequentially on the NBC, CBS, Lifetime, and Spike networks. The show had star-powered hosts over the years (Robert Stack, Dennis Farina, Raymond Burr, Karl Malden, and Virginia Madsen). The later episodes are widely available as reruns. Some of the stories were about unsolved crimes and missing persons – not so much about urban legends – but others were about conspiracy theories, UFOs, and other paranormal phenomena that often related to urban legends. The show used actors to reenact the stories, although real-life participants also appeared. The format was similar to these other shows, with four stories that were reenacted then revealed as truth or fiction. In its peak season (1989–90), the show had over 16 million viewers.[43]

The films typically had a different structure from these television programs.[44] The films either provided a full-length account of a single urban legend or offered

---

[40] "MythBusters", *Wikipedia*, accessed 26 June 2018.

[41] "Mostly True Stories?: Urban Legends Revealed", *Wikipedia*, accessed 26 June 2018.

[42] "Beyond Belief: Fact or Fiction", *Wikipedia*, accessed 26 June 2018.

[43] "Unsolved Mysteries", *Wikipedia*, accessed 26 June 2018.

[44] We considered why there may have been so many of these movies in the 1990s and early 2000s, but we did not come to any strong conclusions. As discussed in Chap. 2, urban legends are cautionary tales that appear in times when people are stressed by the dangerous and complex world in which they live. However, these movies had occasionally appeared as early as the 1970s, so it is hard to pin down some particular circumstances that occasioned these films. Film scholar David Bordwell said of these films: the decision to produce these films had nothing to do with public interest or demand, more what producers thought would be interesting or would work, based on their own interest and the fact that they were cheap to produce; these films were mostly B-class movies of interest largely to teenagers, and did not generate large audiences; these films were aimed at the video market and later DVDs, which were more niche markets; these filmed stopped being made when video and DVD markets began to go away, replaced by streaming video; trends changed too, and the film industry is trendy; and there was no grass roots movement or culture demanding these flicks (Private communication to Cortada, November 2018).

up multiple urban legends in the same movie, meshed together or serially told but still having an overarching narrative drive. Table 4.5 provides a list of some of the films inspired by urban legends.[45] The movie *Urban Legend* was of such interest to ISCLR that it was shown, shortly after it came out, at one of the annual meetings. In writing to announce his new encyclopedia of urban legends, Jan Brunvand mentioned this movie:

> Anyone who saw the film *Urban Legend* released last year will remember the library scene. The beautiful student Natalie (played by Alicia Witt) suspects that recent campus deaths and disappearances were inspired by urban legends, the same kind of stories she is studying in a folklore class at New England's Pendleton College. Natalie goes to the college library to consult the ultimate reference work on the subject, a hefty tome titled *Encyclopedia of Urban Legends*. There she finds the proof she is seeking – an illustrated description of the latest killing.
>
> No such reference work exists, of course, but one is in preparation.[46] (Brunvand 2000)

The two sequels to *Urban Legend* not only had a story line based on urban legends, but like the original movie in the series also had an aspect of studying urban legends. Whereas in the original movie the student was taking a folklore class and consulted an urban legend encyclopedia in the library to help figure out what is going on in her life, in *Urban Legend: Final Cut* (2000) the main character is preparing a film on urban legends for her thesis; and in *Urban Legend: Bloody Mary* (2005) the story line involves artwork about urban legends. There is a similar conceit in the earlier film *Candyman* (1992), in which a principal character in the movie is writing a thesis about urban legends.

We saw in Chap. 2 how the involvement of the long-time members of AFU was driven by the entertainment value of scrutinizing urban legends and reinforcement of their belief in practical rationality as a value for living in a complex world. This same entertainment value and reinforcement of practical rationality is apparent in these truth or fiction television shows. The television audience does not care particularly about whether the stories that are being presented to them are actually true or false; in fact, most viewers probably did not know what stories were going to be presented to them. The urban-legend themed movies also have entertainment value, of course, but they also share a feature with the academic study of contemporary legends – namely, that these movies are a way of expressing deep-rooted cultural fears. Indeed, the academics were keenly interested in these movies.

---

[45] Since 2001, there has been an increasing reaction in film to the 9/11 terrorist attacks. See, for example, Connolly (2009); Feblowitz (2009); Melnick (2009).

[46] Brunvand (2000). The encyclopedia he was discussing is: Brunvand (1996).

Table 4.5 Movies inspired by urban legends (Sample)[a]

Year	Movie name	Urban legend
1971	Foster's Release	Babysitter and Man Upstairs
1972	No Deposit, No Return	Baby Roast
1973	The Wicker Man	Wicker Man
1974	Buster and Billie	The Hook
1975	Night Moves	Alligator in the Sewer
1977	The Sitter	Babysitter and Man Upstairs
1979	When a Stranger Calls	Babysitter and Man Upstairs
1980	Alligator	Alligator in the Sewer
1983	Nightmares	Backseat Trouble
1985	Mr. Wrong	Backseat Trouble
1987	Adventures in Babysitting	Babysitter and Man Upstairs
1989	Mystery Train	Vanishing Hitchhiker
1990	Blood Salvage	Stolen Body Parts
1990	The Krays	Smiley Gang
1990	Lisa	Babysitter and Man Upstairs
1992	Candyman	Candyman
1992	The Harvest	Stolen Body Parts
1995	The Babysitter	Babysitter and Man Upstairs
1995	Candyman: Farewell to the Flesh	Candyman
1995	The Donor	Stolen Body Parts
1997	Campfire Tales	Multiple
1997	I Know What You Did Last Summer	The Hook Man
1998	Dead Man on Campus	Roommate Suicide Means 4.0 GPA
1998	Dead Man's Curve	Roommate Suicide Means 4.0 GPA
1998	Urban Legend	Multiple
1999	Allure	Stolen Body Parts
1999	Candyman: Day of the Dead	Candyman
1999	Lover's Lane	The Hook
2000	The Chromium Hook	The Hook
2000	The Hook-Armed Man	The Hook
2000	Urban Legend: Final Cut	Multiple
2000	Urbania	Multiple
2000	Urban Mythology	Multiple
2003	Spare Parts	Stolen Body Parts
2004	The Call	Babysitter and Man Upstairs
2004	Preheat to 425 [degrees]	Baby Roast
2005	Urban Legend: Bloody Mary	Multiple
2006	When a Stranger Calls	Babysitter and Man Upstairs
2013	The Harvest	Organ Thefts
2015	The Curve	Dead Man's Curve

[a]The main sources for this table are Koven (2007); and Bennett and Smith (2007). Bennett and Smith has a more complete list of films that are based on urban legends, as well as a list of fiction based on urban legends

4.6   Conclusions

People who have developed, maintained, and contributed content to websites that debunk urban legends either as a hobby or to serve the public have become frustrated with the explosion of fake information online over the past decade. For some, it is not as much fun anymore. Moreover, the quantity of fake information appearing online makes their job harder. No longer is the role of the debunker as universally appreciated as it once was, inasmuch as the creators of the fake facts and the political partisans who many of the fake fact-makers serve attack the debunkers because the fake facts support their political worldviews. It is a pattern of behavior, criticism, and weariness as old as the United States.

In some cases, such as the urban legend pages on AOL or even AFU, the sites have fallen out of favor because of changes in the technologies in online communication. But in other cases, people have simply abandoned their debunking exercises because it is not worth the effort. Consider the popular urban legend debunking website hoaxbusters.org.[47] It closed down at the end of 2016 not because there was no longer any need for the site but because the nature of the problem had changed. The Hoaxbusters.org website pointed its visitors to eight other "excellent hoax-debunking websites": Hoax-Slayer, TruthorFiction.com, FactCheck.org, PolitiFact, *Washington Post* Fact Checker, ThatsFake.com, ThatsNonsense.com. and snopes. But as Hoaxbuster.org explains in its final online message:

> After 17 years of hoax-busting, the time has come to call it a day. It has been our pleasure to serve you since 1999, and we are honored to have been one of the trusted sources that you chose for hoax debunking. But all good things must come to an end, and Hoax Busters is no different. As of January 1, 2017, we are officially retired. The tenor of hoaxes has changed through the years. These days, it's all about conspiracy theories and political misinformation. Those types of hoaxes are spread by folks whose only interest is in reading news that conforms to their point of view. No matter the actual facts, people will believe what they want to, and truth is irrelevant. Walter Quattrociocchi, the head of the Laboratory of Computational Social Science at IMT Lucca in Italy, has spent several years studying how conspiracy theories and misinformation spread online. He explained that institutional distrust is so high, and cognitive bias so strong, that the people who fall for hoax news stories are frequently only interested in consuming information that conforms with their views – even when it's demonstrably fake.[48]

In these various comments one senses a weariness, almost the exhaustion that comes from shoveling sand against the sea – too much fake news popping up all over the Internet, too many sole misinformers competing for people's attention with large institutional players in the private and public sectors promoting content in support of their parochial interests with little or no respect for truth. That this had been a problem in the pre-Internet era was irrelevant, because it was a problem now.

---

[47] This is not to be confused with hoaxbusters.ciac.org, profiled earlier. It should also not be confused with HoaxBustersCall.com: Conspiracy or Just Theory?, which is a blog and an Internet radio show available on Stitcher.

[48] http://www.hoaxbusters.org accessed 31 May 2018.

Consumers of such misinformation, rumors, and hoaxes were just learning that not all content was truthful – it is a learning process very much still underway.

This chapter shows how seemingly unconnected phenomena appearing at the same time as AFU and snopes – the creation of a new academic discipline by sociologists and folklorists, the appearance of a new genre of grade-B horror films, and efforts to protect people from computer viruses – are all actually connected through an interest in scrutiny to make sense of and live safely in a complex, dangerous world. It is important to reiterate a point made in Chap. 1: this scrutiny appears in a variety of forms; sometimes it is about ascertaining the veracity of a claim, sometimes about understanding the cultural meaning of an urban legend (whether or not it is true), and sometimes it is about using truth and falsity as a tool to understand if a website or email is authentic when it promises the viewer some reward by clicking on a site (or whether it will lead one into the danger of computer viruses).

The next chapter discusses this modern era of political misinformation, which so turned off the authors of the Hoaxbuster.org website and that represented the new flood of fake information. As we will see, this era of political misinformation has been accompanied by numerous efforts at political fact-checking. They are yet another contemporaneous – indeed, the most well-known – of these examples of scrutinizing in a modern, complex, dangerous world; and in this case, veracity is a primary consideration.

References

Anon. 1996. AOL Unveils The Hub. *Advertising Age,* March 25. http://adage.com/article/news/aol-unveils-hub/1728/. Accessed 24 July 2017.

———. 2009. Willie Lynch Letter: The Making of a Slave. *The Final Call,* May 22. http://www.finalcall.com/artman/publish/Perspectives_1/Willie_Lynch_letter_The_Making_of_a_Slave.shtml. Accessed 25 July 2017.

Bacon, Jasen. 2011. *The Digital Folklore Project: Tracking the Oral Tradition on the World Wide Web*. Electronic Theses and Dissertations. Paper 1398. http://dc.etsu.edu/etd/1398.

Bennett, Gillian, and Paul Smith. 1988. Introduction. In *Monsters with Iron Teeth: Perspectives on Contemporary Legend III*. Sheffield: Sheffield Academic Press.

———. 1989. *The Questing Beast: Perspectives on Contemporary Legend IV*. Sheffield: Sheffield Academic Press.

———. 1993. *Contemporary Legend: A Folklore Bibliography*. New York: Garland.

———. 2007. *Urban Legends: A Collection of International Tall Tales and Terrors*. Westport: Greenwood Press.

Bennett, Gillian, Paul Smith, and J.D.A. Widdowson, eds. 1987. *Persepctives on Contemporary Legend II*. Sheffield: CECTAL/Sheffield Academic Press.

Brunvand, Jan Harold, ed. 1996. *American Folklore: An Encyclopedia*. New York: Garland.

———. 2000. An Encyclopedia of Urban Legends. *FOAFTale News*, No. 47, October. http://www.folklore.ee/FOAFtale/ftn47.htm#abstracts. Accessed 22 June 2018.

———. 2004. *'Urban Legend' – Still Booming, Despite Vanishing*. Paper Abstract, 22nd Annual Perspectives on Contemporary Legend Conference, Aberystwyth, Wales, in *FOAFTale News*, No. 59, August. http://www.folklore.ee/FOAFtale/ftn59.htm. Accessed 22 June 2018.

———. 2005. 'Urban Legend' from 1925. *FOAFTale News*, No. 62, August. http://www.folklore.ee/FOAFtale/ftn62.htm. Accessed 22 June 2018.

Buck, Stephanie. 2017. In the 90s, Bizarre Rumors and Urban Legends Spread to the Masses via Fax Machine. *Timeline,* February 8. https://timeline.com/urban-legend-fax-machine-e3ec2ce15d7b. Accessed 25 July 2017.

Cheng, Kipp. 1996. Website Review: AOL's Urban Legends. *Entertainment Weekly,* November 29. http://ew.com/article/1996/11/29/website-review-aols-urban-legends/. Accessed 24 July 2017.

Christensen, Brett. 2016. *Badtimes Spoof Makes Fun of Old 'Good Times' Virus Hoax,* ed. Hoax-Slayer, July 16. https://www.hoax-slayer.net/badtimes-spoof-makes-fun-old-good-times-virus-hoax/. Accessed 10 Apr 2019.

CIAC and the U.S. Department of Energy. 2000. *Welcome to the New CIAC Hoax Pages,* November 13. https://web.archive.org/web/20001202035600/http://hoaxbusters.ciac.org:80/HoaxBustersHome.html. Accessed 11 June 2018.

CNN financial news staff. 2000. Primedia Buys About.com. *CNN Money,* October 30. http://money.cnn.com/2000/10/30/deals/pri/. Accessed 28 July 2017.

Connolly, Matthew James. 2009. *Reframing the Disaster Genre in a Post-9/11 World.* Honors Thesis, Film Studies, Wesleyan University. http://wesscholar.wesleyan.edu/cgi/viewcontent.cgi?article=1237&context=etd_hon_theses. Accessed 29 June 2018.

Cooper, Deborrah. 2010. The Urban Legend that Is the Willie Lynch Letter. *Surviving Dating! Blog,* November 16. http://survivingdating.com/the-urban-legend-that-is-the-willie-lynch-letter. Accessed 25 July 2017.

Dale, Rodney. 1978. *The Tumour in the Whale.* London: Duckworth.

Degh, Linda. 1994. *American Folklore and the Mass Media.* Bloomington: Indiana University Press.

Degh, Linda, and Andrew Vazsonyi. 1983. Does the Word 'Dog' Bite? Ostensive Action: A Means of Legend Telling. *Journal of Folklore Research* 20: 5–34.

Delio, Michelle. 2001. The Man Who Debunks Virus Myths. *Wired,* August 6. https://www.wired.com/2001/08/the-man-who-debunks-virus-myths/. Accessed 19 June 2018.

Eco, Umberto. 1976. *A Theory of Semiotics.* Bloomington: Indiana University Press.

Edwards, Ben J. 2014. Where Online Services Go When They Die: Rebuilding Prodigy, One Screen at a Time. *The Atlantic,* July 12. https://www.theatlantic.com/technology/archive/2014/07/where-online-services-go-when-they-die/374099/. Accessed 27 July 2017.

Eisenberg, Ted, et al. 1989. The Cornell Commission: On Morris and the Work. *Communications of the ACM* 32(6, June): 706–709.

Ellis, Bill. 1989. Needling Whitey: The New York City Pin-Prick Incidents as Ostention. *FOAFTale News*, No. 16, December. http://www.folklore.ee/FOAFtale/ftn16.pdf. Accessed 21 June 2018.

———. 1991a. 'Contemporary Legend' – Cracks or Breakthroughs. *Folklore* 102: 183–186.

———. 1991b. Contemporary Legend – 'The Pale Cast of Thought?' *FOAFTale News,* No. 21, March. http://www.folklore.ee/FOAFtale/ftn21.pdf. Accessed 21 June 2018.

———. 2002. The Roots of 'Perspectives on Contemporary Legend': 'Urban Myths' in the 1960 Rhodes-Livingstone Institute for Social Research Conference. Abstract from the Perspectives on Contemporary Legend 2002 Conference, Sheffield, in *FOAFtale News*, No. 53, December 2002, http://www.folklore.ee/FOAFtale/ftn53.htm#abstracts. Accessed 22 June 2018.

Ellis, Bill, and Alan E. Mays. 1994. Art Linkletter and the Contemporary Legend: A Bibliographical Essay. *FOAFtale News* No. 33–34, June. http://www.folklore.ee/FOAFtale/ftn33-34.pdf. Accessed 22 June 2018.

Feblowitz, Joshua C. 2009. The Hero We Create: 9/11 and the Reinvention of Batman. *Inquiries* 1(12): 1. http://www.inquiriesjournal.com/articles/104/the-hero-we-create-911-the-reinvention-of-batman. Accessed 29 June 2018.

Festa, Paul. 1998. AOL's Hub Closes. *CNET,* March 24. https://www.cnet.com/news/aols-hub-closes/. Accessed 24 July 2017.

Fine, Gary Alan. 1991. Redemption Rumors and the Power of Ostension. *Journal of American Folklore* 104: 179–181.

Fitzgerald, F. Scott. 1925. *The Great Gatsby.* New York: Charles Scribners' Sons.

Fleming, Eric. 1999. Miningco.com Changes Name to About.com. *ZDNet,* May 17. http://www.zdnet.com/article/miningco-com-changes-name-to-about-com/. Accessed 28 July 2017.

Fraser, Powell. 2003. How Some Spammers Get Your Email. *CNN.com International,* September 2. http://edition.cnn.com/2003/TECH/internet/09/01/spam.chainletter/index.html. Accessed 31 May 2018.

Garun, Natt. 2017. About.com, the General Interest Site Even Its Own CEO Doesn't Care for, Is Going Away. *The Verge,* April 26. https://www.theverge.com/2017/4/26/15433810/about-com-shut-down-rebrand. Accessed 28 July 2017.

Glanton, Dahleen. 2006. E-Mails Give Legs to Myths. *Seattle Times,* January 9. https://www.seattletimes.com/business/e-mails-give-legs-to-myths/. Accessed 31 May 2018.

Goff, Leslie. 1998. Hoax on You. *Computerworld,* May 25, 71–73

gordonf. 2004. Addictive Update Model. *Everything2,* December 20. https://everything2.com/title/addictive+update+model. Accessed 19 June 2018.

Gosztola, Kevin. 2007. Introducing ... Lew Koch! *Shadowproof,* May 14. https://shadowproof.com/2007/05/14/introducing-lew-koch/. Accessed 19 June 2018.

Hand, Wayland D. 1971. *American Folk Legend: A Symposium.* Berkeley: University of California Press.

Henke, Elissa. 2010. Minutes of ISCLR's Annual General Meeting, June 30, 2010. *FOAFTale News,* No. 76, September. http://www.folklore.ee/FOAFtale/ftn76.htm. Accessed 22 June 2018.

Hobbs, Sandy. 2010. Scholarly Use of the Term 'Urban Legend'. *FOAFTale News,* No. 74, January. http://www.folklore.ee/FOAFtale/ftn74.htm. Accessed 22 June 2018.

Howitt, Doran. 1984. The Source Keeps Trying: Does America Need an Information Utility? *InfoWorld,* November 5, 59–64. https://books.google.com/books?id=oC4EAAAAMBAJ&pg=PA59#v=onepage&q&f=false. Accessed 25 July 2017.

InformationWeek News Staff. 1999. Y2K Shuts Down Prodigy Classic. *InformationWeek,* January 25. http://www.informationweek.com/y2k-shuts-down-prodigy-classic/d/d-id/1006720. Accessed 27 July 2017.

Ingram, Mathew. 2017. About.com Has a New Name and a New Strategy. *Fortune,* May 2. http://fortune.com/2017/05/02/about-reinvents-dotdash/. Accessed 28 July 2018.

Jason, Heda. 1990. Contemporary Legend' – To Be or Not to Be? *Folklore* 101: 221–223.

Koven, Mikel. 2007. *Film, Folklore, and Urban Legends.* Lanham: Scarecrow Press.

Kushner, David. 1997. Drilling Begins at the Mining Company. *Wired News,* April 21. https://web.archive.org/web/20040910143418/http://www.wired.com/news/culture/0,1284,3272,00.html. Accessed 28 July 2017.

Kvartic, Ambroz. 2013. ISCLR Conference – Some Quick Impressions. *FOAFTale News,* No. 80, January. http://www.folklore.ee/FOAFtale/ftn80.pdf. Accessed 22 June 2018.

Melnick, Jeffrey. 2009. *9/11 Culture.* New York: Wiley-Blackwell.

Smith, George. 1994. *The Virus Creation Labs: A Journey into the Underground.* Show Low: American Eagle Publications.

MonsieurEvil. 2003. VMyths.com in Trouble. *Subnautica,* July. https://forums.unknownworlds.com/discussion/37857/vmyths-com-in-trouble. Accessed 19 June 2018.

Moravec, Mark. 1995. Perspectives on Contemporary Legend: The Paris Conference. *FOAFtale News* No. 37, June. http://www.folklore.ee/FOAFtale/ftn37.htm#paris. Accessed 25 June 2018.

Moser, Whet. 2007. Lewis Koch Joins the Blogosphere. *Chicago Reader,* May 17. https://www.chicagoreader.com/Bleader/archives/2007/05/17/lewis-koch-joins-the-blogosphere. Accessed 19 June 2018.

Moyer, Heather. 2001. E-mail Hoaxes Abound. *Disaster News Network,* October 15. https://www.disasternews.net/news/article.php?articleid=968&printthis=1. Accessed 31 May 2018.

Online Timetable. n.d. *The Source.* http://iml.jou.ufl.edu/CARLSON/history/the_source.htm. Accessed 25 July 2017.

Rankin, Joy Lisi. 2018. *A People's History of Computing in the United States.* Cambridge, MA: Harvard University Press.

Shapiro, Eben. 1990. THE MEDIA BUSINESS; New Features are Planned by Prodigy. *The New York Times,* September 6. http://www.nytimes.com/1990/09/06/business/the-media-business-new-features-are-planned-by-prodigy.html. Accessed 27 July 2017.

Smith, Paul, ed. 1984. *Perspectives on Contemporary Legend: Proceedings of the Conference on Contemporary Legend.* Sheffield, July 1982. Sheffield: Centre for English Cultural Tradition and Language Publications.

———. 1999. Definitional Characteristics of the Contemporary Legend. *FOAFtale News,* No. 44, May. http://www.folklore.ee/FOAFtale/ftn44.htm. Accessed 25 June 2018.

———. 2002. Teaching Folklore 3612: Urban Legend. *FOAFtale News,* No. 53, December. http://www.folklore.ee/FOAFtale/ftn53.htm#abstracts. Accessed 22 June 2018.

Smith, George. 2008. Chinese Cybermen Said to Be Impurifying Our Precious Bodily Fluids: Cyber-Arclight Strikes Eyed as Remedy. *Dick Destiny,* June 9. http://www.dickdestiny.com/blog/2008/06/chinese-cybermen-said-to-be-impurifying.html. Accessed 19 June 2018.

Sones, Bill, and Rich Sones. 2005. *Can a Guy Get Pregnant?: Scientific Answers to Everyday (And Not So Everyday) Questions.* New York: Pi Press.

Sowa, Logan. 2016. Terrifying Urban Legends from Across the Country. *AOL,* October 19. https://www.aol.com/article/lifestyle/2016/10/19/terrifying-urban-legends-from-across-the-country/21587137/. Accessed 24 July 2017.

Stratford, Suzanne. 2016. Urban Legend or Satanic Site? A Look at 'Satan's Hollow'. *AOL,* February 9. https://www.aol.com/article/2016/02/09/urban-legend-or-satanic-site-a-look-at-satans-hollow/21310166/. Accessed 24 July 2017.

VMyths. 2008. *Website Charter,* November 10. http://vmyths.com/about/website-charter/. Accessed 19 June 2018.

Webb, J. A. 1989, July. CompuServe Purchases the Source. *Information Today* 6(7): 1.

Weisbaum, Herb. 2007. Urban Legends Outlawed…April Fools. *NBC News,* 4/1/2007. http://www.nbcnews.com/id/17798063/ns/business-consumer_news/t/urban-legends-outlawed-april-fools/#.WXuFFa3MwrU. Accessed 28 July 2017.

Widdowson, John. 2002. From Small Beginnings: The Gestation of Contemporary Legend Research, Abstract from the Perspectives on Contemporary Legend 2002 Conference, Sheffield, in *FOAFtale News, No. 53,* December 2002. http://www.folklore.ee/FOAFtale/ftn53.htm#abstracts. Accessed 22 June 2018.

Wikipedia. n.d. *Computer Incident Advisory Capability.* Accessed 31 May 2018.

Wired staff. 1997. Icon Turns to AOL's Hub for Exposure. *Wired,* July 1. https://www.wired.com/1997/07/icon-turns-to-aols-hub-for-exposure/. Accessed 24 July 2017.

Woollcott, Alexander. 1934. *While Rome Burns.* New York: Grosset & Dunlap.

Chapter 5
Recent Political Fact-Checking

Truthfulness has never been counted among the political virtues.
– Hannah Arendt (Arendt (1971), quoted in Graves (2016))

The fact is that President Trump lies not only prolifically and shamelessly, but in a different way than previous presidents and national politicians. They may spin the truth, bend it, or break it, but they pay homage to it and regard it as a boundary. Trump's approach is entirely different. It was no coincidence that one of his first actions after taking the oath of office was to ... put the press and public on notice that he intended to bully his staff, bully the media, and bully the truth.
– Jonathan Rauch (Rauch 2018)

Over the past decade, American politics has been increasingly riddled with fake facts spread rapidly and widely across the Internet. One might date this as beginning with the 2008 Presidential election campaign, although this activity does have a long pre-history. It is often easier to spread fake facts than true ones because fake facts can be crafted that resonate with the desires and fears of various groups of citizens. We saw in Chap. 2 how snopes.com was transformed from being a hobbyist site that evaluated urban legends and scientific curiosities to one engaged in the serious activity of political fact-checking.

One of the responses from the legitimate media has been the emergence of fact-checking, to identify what is fake and what is real in online accounts. As one observer of these developments has argued:

[I]t's helpful to think about why "fact-checking" has emerged now. I'd argue that it's a response to many journalists' perception that they are ever more outgunned by the increasing volume and sophistication of professional political communication. The fact-check is a tool with which reporters can rescue themselves from oblivion. And the morally freighted language invoked by full-time fact-checkers – true and false, fact and lie – is a weapon, to be wielded by journalists with authority against other, presumably less trustworthy types who make political claims. (Marx 2012)

© Springer Nature Switzerland AG 2019
W. Aspray, J. W. Cortada, *From Urban Legends to Political Fact-Checking*,
History of Computing, https://doi.org/10.1007/978-3-030-22952-8_5

Table 5.1 Pro-trust initiatives, including political fact-checking organizations (sample)

2003	FactCheck.org	University of Pennsylvania Annenberg Public Policy Center
2007	PolitiFact	St. Petersburg (Florida) Times
2007	The Fact Checker	Washington Post
2013	PunditFact	Ford Foundation and the Democracy Fund
2014	Accountability Journalism Program	American Press Institute
2015	Polygraph.info	Voice of America and Radio Free Europe/Radio Liberty
2015	The Trust Project	University of Santa Clara Markkula Center for Applied Ethics
2016	Trusting News	Reynolds Journalism Institute, University of Missouri
2016	Media Manipulation Initiative	Data & Society Research Institute
2017	Fact-checker	The Weekly Standard
2017	News Integrity Initiative	CUNY Craig Newmark Graduate School of Journalism
2017	Deepnews.ai (formerly the news quality scoring project)	Initiated at Stanford University with funding from the Knight Foundation, Democracy Fund, and Rita Allen Foundation
2017	Duke Tech & Check Cooperative	Knight Foundation, the Facebook Journalism Project, and the Craig Newmark Foundation
2017	WikiTribune	Jimmy Wales for-profit corporation
2018	NewsGuard	Steven Brill and Gordon Crovitz entrepreneurial venture
2018	Journalism Trust Initiative	Reporters Without Borders (RSF)
2018	Our.News	Independent start-up

What is particularly interesting to observe is how many of these fact-checking organizations emerged during the first two decades of the new century. See Table 5.1. The leading ones are well-funded and staffed with experienced journalists and other researchers. Their existence is as much a testament to the continuing desire for accurate facts in American society as they are a reminder that so much disinformation and so many falsehoods float through the Internet and spill over into print. These Internet-based truth-seekers represent a new and growing genre in the digital world.[1] They may be the vanguard of a new generation of fact seekers because one of the byproducts of the American elections of 2016 and 2018 has been a rising surge of interest in regulating social media platforms precisely to block falsehoods, e.g. to prevent Facebook, in its Trending News section, from posting false claims such as the US government was behind the 9/11 attacks.[2] We suspect that the websites described below point to a rapidly expanding national response to the essential American belief that an accurately informed citizenry is crucial to the functioning of the US democracy.

---

[1] A trend already identified by American journalists in their industry as early as 2007, Silverman (2007).

[2] See, for example, Minor (2016); Caitlin Dewey (2016).

This chapter is organized roughly chronologically, with US presidential elections as an important delineator between waves of fact-checking organizations. We begin with a brief pre-history of US political fact-checking. We end that section with the creation of FactCheck.org in advance of the George W. Bush – John Kerry presidential election in 2004. At that time, it was still not entirely clear how big an issue fake facts was going to be. The next section covers the emergence of the *Washington Post* fact checker and PolitiFact in response to the heightened concern over fake facts in the coming 2008 election between Barack Obama and John McCain. PolitiFact and its sister site, PunditFact, gets significantly more attention in this chapter than any other fact-checking organization because it is considered one of the most reliable arbitrators of the truth and as a consequence quickly became one of the primary sources of verifiable facts for academics, media, and politicos themselves when facts were doubted. The next section covers the fact-checking organizations created in advance of the 2016 election between Donald Trump and Hillary Clinton.[3] These include the Accountability Journalism Program of the American Press Institute, Polygraph.info of Voice of America, and the Trusting News initiative of the University of Missouri journalism school. When it was recognized how significant a force fake facts, including Russian interference using fake facts as a weapon, may have been in shaping the outcome of the 2016 election, there was a tsunami of new political fact-checking organizations: the Media Manipulation Initiative, *The Weekly Standard* Fact-checker, The Trust Project of the Markkula Center for Applied Ethics, the News Integrity Initiative of the CUNY journalism school, Deepnews.ai started at Stanford University, the Duke Tech & Check Initiative, WikiTribune from Wikipedia founder Jimi Wales, NewsGuard, the international Journalism Trust Initiative, and the for-profit start-up Our.News. It is anticipated that other initiatives will continue to be added.

5.1   Pre-history of Political Fact-Checking in the United States

Major national magazines had begun to build up large fact-checking departments in the 1920s and 1930s (Graves 2016). However, the purpose of these departments was to check the veracity of the stories journalists wrote for the magazines.[4] These departments began to be scaled back in the late twentieth century as the news media came upon increasingly hard times. For example, *Newsweek* closed its fact-checking organization in the 1990s. In contrast, the goal of the political fact-checker today is to check the veracity of the claims made by the politician or the political group. In the twenty-first century, the major broadcast networks, National Public Radio, the

---

[3] While the existing political fact-checking organizations were diligent during the 2012 presidential election between Barack Obama and Mitt Romney, there were few new political fact-checking organizations created in preparation for that election.

[4] Silverman, *Regret The Error,* 275–279.

New York Times, *USA Today*, and Associated Press have invested in fact-checking (Graves 2016). *The Wall Street Journal* was perhaps the only major news organization not making news out of political fact-checking.

Before there was concerted attention to truth in American journalism, one American organization, Internews, founded in 1982 in San Francisco with support from the Kenall Foundation, began to work toward a fair and independent media around the world, addressing at first television and later the Internet.[5] These goals of fairness and independence were at the heart of the later efforts to combat fake facts in American politics and journalism. As the nonprofit describes its mission today:

> Internews works across a range of issues in the fast-moving information and media landscape to reach the most information poor and disadvantaged. From fighting propaganda and corruption to combatting extremism, from protecting a free and open internet to media and data literacy, from strengthening governance and health systems to supporting local media.[6]

Both the hanging-chad brouhaha in Florida during the 2000 presidential election between George W. Bush and Al Gore and the 9/11 terrorist attacks in 2001 led a number of academic journalism organizations to become concerned about deception in the claims of participants in the political process. As the country prepared for the 2004 presidential election, concern began to mount about this issue. FactCheck. org, which is a project of the University of Pennsylvania's Annenberg Public Policy Center, was founded in 2003. It describes itself as:

> a nonpartisan, nonprofit "consumer advocate" for voters that aims to reduce the level of deception and confusion in U.S. politics. We monitor the factual accuracy of what is said by major U.S. political players in the form of TV ads, debates, speeches, interviews and news releases. Our goal is to apply the best practices of both journalism and scholarship, and to increase public knowledge and understanding.[7]

---

[5] Here is a brief history of Internews, pulled from its webpage (https://www.internews.org/our-history, accessed 18 December 2018). In 1982, Internews won an Emmy Award for linking by satellite the U.S. Congress with Deputies in the Supreme Soviet. The goal was to bridge the East-West divide using television. In 1990, they moved away from international television programming and began to support non-governmental media in Eastern Europe and the former Soviet Union. In 1993, Internews teams up with the Jerusalem Film Institute to train for news production and election coverage for Palestinian television. In 1994, Internews established bulletin board and email systems, called the Balkan Media Network, to enable independent media networks and individuals to have contact with the outside world while under assault from the Yugoslav National Army. In 1997, Internews began coverage of the UN tribunal for Rwanda. The following year, Internews began to train and support independent radio stations in Indonesia. In 2001, it initiated the Global Internet Policy Initiative. In 2002, Internews began to help create local, professional media in Afghanistan and provide coverage of the AIDS epidemic in Africa. In 2004, Internews created an Earth Journalism Network to help developing countries to cover environmental issues; and in 2009, it established awards for journalistic coverage of climate change. In 2011, Internews provides support to free media in the Middle East, during the Arab Spring uprising. In 2013 it increases its work on Internet freedom and digital security. In 2014, Internews creates programs to grow the independent media in the South Sudan and Somalia. In 2015, Internews established a program to ensure that women's voices could be hear around the world.

[6] https://www.internews.org/key-issues-menu

[7] https://www.FactCheck.org/about/our-mission/, accessed 29 June 2018.

FactCheck.org attempts to monitor Republicans and Democrats equally, systematically monitoring Sunday talk shows, televisions advertisements, C-SPAN videos of campaign rallies and events, presidential remarks, *Congressional Quarterly* transcripts, campaign and official websites and press releases, and questions from readers. This fact-checking site uses primary sources of information and "trustworthy" outside experts, such as the Kaiser Family Foundation, the Tax Policy Center, and the National Conference of State Legislatures. FactCheck.org does not accept funding from unions, partisan organizations, or advocacy groups; and it employs a highly transparent process for disclosing donors. Staff members are mostly experienced journalists, with a college professor or two mixed in. The website contains current stories plus an archive of older stories. In 2017, FactCheck.org partnered with Facebook to enable Facebook users to report stories they believe are fake. This initiative was in response to many complaints about fake political stories appearing on Facebook during the 2016 election (Schaedel 2017).

5.2   Run-Up to the 2008 Presidential Election

There had been highly partisan coverage and spread of misleading stories about hanging chads in the nail-bitingly close 2000 presidential election between George W. Bush and Al Gore, as well as in the coverage of Democratic candidate John Kerry (e.g., the Swift Boat Veterans for Truth commercial) in the 2004 presidential election. There were not many new initiatives in combatting fake news at the time of the 2004 election, however, perhaps because Bush was an incumbent. But with the field open for the 2008 presidential election between Barack Obama and John McCain, there was anticipation that fake news would greatly increase, as it did for example in the stories about Obama's place of birth. We consider three initiatives in this section: the creation of the *Washington Post* Fact Checker, the rise of fact-checking organization PolitiFact, and later the creation of PolitiFact's sister organization, PunditFact. We devote considerable attention in this section to the PolitiFact story because we regard this organization, together with snopes, as the most important developments in the field of political fact-checking.

Award-winning journalist Glenn Kessler runs The Fact Checker for the *Washington Post*. During the 1992 and 1996 presidential campaigns, he fact-checked candidate statements for *Newsday*, where he was chief political correspondent. He moved over to the *Washington Post*, first to cover tax and business policy, then as the national business editor, and later as the chief State Department reporter. At the *Post*, he has covered each presidential election since 2000. His column, The Fact Checker, began as a feature on September 19, 2007 as part of the coverage of the 2008 presidential campaign. It was abandoned after the 2008 election ended, but revived and made a permanent feature of the *Washington Post* on January 11, 2011 as part of the coverage of the 2012 presidential election. As Kessler describes The Fact Checker website:

Table 5.2 The *Washington Post* fact checker rating system[a]

1 Pinocchio	Some shading of the facts. Selective telling of the truth. Some omissions and exaggerations, but no outright falsehoods. (You could view this as "mostly true.")
2 Pinocchios	Significant omissions and/or exaggerations. Some factual error may be involved but not necessarily. A politician can create a false, misleading impression by playing with words and using legalistic language that means little to ordinary people. (Similar to "half true.")
3 Pinocchios	Significant factual error and/or obvious contradictions. This gets into the realm of "mostly false." But it could include statements which are technically correct (such as based on official government data) but are so taken out of context as to be very misleading. The line between Two and Three can be bit fuzzy and we do not award half-Pinocchios. So, we strive to explain the factors that tipped us toward a Three.
4 Pinocchios	Whoppers

[a]Source: Kessler (2013b). The descriptions in the right-hand column are verbatim

> The purpose of this website, and an accompanying column in the Sunday print edition of *The Washington Post*, is to "truth squad" the statements of political figures regarding issues of great importance, be they national, international or local. It's a big world out there, and so we rely on readers to ask questions and point out statements that need to be checked.
> But we are not limited to political charges or coun111charges. We also seek to explain difficult issues, provide missing context and provide analysis and explanation of various "code words" used by politicians, diplomats and others to obscure or shade the truth. (Kessler 2013a)

Approximately half of the facts that are checked are initiated by an inquiry from a reader. As Kessler notes, "Our view is that a fact check is never really finished, so the rating can be revised after we obtain new information that changes the factual basis for our original ruling" (Kessler 2013a, b). The focus is on verifiable facts, not opinions. The goal is to be "dispassionate and non-partisan". The policy is for no one associated with The Fact Checker to be engaged in political advocacy or making campaign contributions to any political cause or party. However, as Kessler observes: "But we also fact check what matters – and what matters are people in power. When one political party controls the White House and both houses of Congress, it is only natural that the fact checks might appear to be heavily focused on one side of the political spectrum" (Kessler 2013a, b).

The Fact Checker has a rating system as described in Table 5.2. Kessler has built on the system of giving between one and four Pinocchio ratings to false or misleading statements, to include an upside-down Pinocchio for a "statement that represents a clear but unacknowledged 'flip-flop' from a previously-held position" and a Bottomless Pinocchio for a claim that was awarded three or four Pinocchios when it was first evaluated and which has been repeated by the speaker at least 20 times.[8]

The *Washington Post* is widely regarded as politically liberal in its editorial policy, and so its fact-checking has generally been an object of attack from

---

[8] Kessler 2013. For further information about the rating system, see the transcript of an interview of Kessler conducted by Brian Lamb for C-SPAN on December 22, 2011, https://www.c-span.org/video/?303324-1/qa-glenn-kessler

conservatives about its political bias.[9] The citizens watchdog nonprofit organization Accuracy in Media has covered approximately 30 stories about the accuracy of The Fact Checker ratings, complaints about these ratings, and misuse of the ratings by others.[10] We discuss political bias later in this chapter, so we will not provide a detailed analysis of the bias claims here. Instead, we turn next to our major profile of a fact-checking site – of PolitiFact and its sister site PunditFact.

The major development in political fact-checking for the 2008 election was the creation of PolitiFact. It is the largest fact-checking organization in the United States. Its professional staff – made up of well-established editors and journalists – includes an editor, executive director, deputy editor, senior correspondent, six staff writers, an audience engagement fellow, associate editor, and regional editors and writers for Texas, Wisconsin, New York, California, and North Carolina. It was founded in 2007 as an election-year project of the *St. Petersburg Times* (today known as the *Tampa Bay Times*). Bill Adair, who was the Washington bureau chief for the newspaper, proposed the idea for PolitiFact. Since 2008, the non-profit Poynter Institute has owned PolitiFact for Media Studies. The Poynter family had owned the *St. Petersburg Times* throughout most of the twentieth century, while Nelson Poynter had also founded the *Congressional Quarterly*, the principal publication covering news of Congress.[11] PolitiFact is widely acclaimed, e.g. winner of the 2009 Pulitzer Prize for National Reporting.

PolitiFact runs a partner website, PunditFact, which conducts fact-checking on pundits, bloggers, and hosts and guests on talk shows. It was created in 2013 with funding from the Ford Foundation and the Democracy Fund (Graves 2016). PolitiFact has also franchised its brand and methodology to major newspapers to cover state politics; as of 2015, 14 states had franchises (Graves 2016). Some well-known PolitiFact partners include the *Atlanta Journal-Constitution* and *Austin American-Statesman* (newspapers that are part of Cox Media Group), the *Milwaukee Journal Sentinel* (part of Gannett), Capital Public Radio, and Billy Penn mobile-first news.

While the original funding came from the *St. Petersburg Times*, today funding for PolitiFact comes from content partnerships, online advertising, grants, and

---

[9] As illustration of these comments, see for example, "Washington Post Caught Red Handed Peddling Anti-Trump Fake News," editorial, *Investor's Business Daily*, September 18, 2018, https://www.investors.com/politics/editorials/washington-post-fake-news-passports-media-bias/ (accessed 20 December 2018); or "110 Examples of Post Misreporting," eyeonthepost.org, covering reporting during the period May to October 2002., http://www.eyeonthepost.org/110examples.html (accessed 20 December 2018). As for whether the Post is liberal and how liberal, the website Media Bias/Fact Check evaluated the Washington Post and determined it has a left-center bias (which is more moderate than their ratings of left bias or extreme left bias) and that it rates High in factual reporting on account of its use of proper sources. (https://mediabiasfactcheck.com/washington-post/, accessed 20 December 2018).

[10] "Articles relating to: Washington Post fact checker", Accuracy in Media, https://www.aim.org/tag/washington-post-fact-checker/ (accessed 20 December 2018).

[11] Much of the basic organizational information in this and the next several paragraphs is taken from the Wikipedia article about PolitiFact (accessed the week of June 25, 2018). Additional information is taken from Graves (2016).

individual donations from readers. The management discloses donations or grants in excess of $1000 and does not accept grants from anonymous sources, political parties, elected officials, candidates for elected office, or any other source that is perceived to have a conflict of interest.

PolitiFact communicates its findings on its website through the so-called Truth-O-Meter.[12] The website has rendered a judgment on approximately 15,000 claims and positions of politicians and political groups. The Truth-O-Meter gives a judgment on political claims of either true, mostly true, half true, false, mostly false, or "pants on fire" – the latter meaning that "the statement is not accurate and makes a ridiculous claim".[13] PolitiFact also evaluates political positions taken by individuals or organizations, using the following categories: no flip, half flip, or full flop. The summary judgment given by the Truth-O-Meter includes the name of the person making the claim, the date, and a quotation from the claimant that is typically 20 words or less and is backed up with a story explaining PolitiFact's reasoning and evidence, including sources consulted and names of individuals interviewed, restricting individuals to on-the-record comments.[14]

The Truth-O-Meter judgments are organized in several ways for ease of finding material on the website: chronologically, by topic (e.g. climate change or immigration), and by person who makes the claim. President Trump, former President Obama, Vice President Pence, and the 2018 Congressional leadership (Mitch McConnell, Paul Ryan, Nancy Pelosi, and Chuck Schumer) all have special folders to make it easier to find their claims. The website also has a special file on chain emails that express political viewpoints. These chain emails PolitiFact has found to be particularly untruthful. The Truth-O-Meter indicates that for chain emails expressing political views, 3% are true, 3% mostly true, 4% half true, 11% mostly false, 24% false, and 59% are deemed "pants on fire".

It is interesting to compare the campaign promises made by Presidents Obama and Trump. Of 533 campaign promises made by Obama, PolitiFact found that 48% were kept, 27% involved a compromise, and 24% were broken. On evaluating President Trump's campaign promises, PolitiFact found 11% kept, 7% fulfilled with compromise, 42% in the works, 33% stalled, and 7% broken, so far.[15]

With a task as sensitive as making judgments about political claims, PolitiFact of course has to take great care with its methods and practices. It has codified them in

---

[12] There is a thoughtful analysis of the Truth-O-Meter in Graves (2016), Chapter 5, "Operating the Truth-O-Meter." Also see Chap. 2 of the same book on the epistemology of political fact-checking.

[13] politifact.com, accessed 27 June 2018.

[14] PolitiFact maintains a spreadsheet of political claims, known as "the buffet", that are candidates for evaluation by the Truth-O-Meter. The process of finding political claims to add to the spreadsheet is known inside PolitiFact as "stocking the buffet" and as "trawling for shrimp". See Graves (2016).

[15] 'Compromise' means to PolitiFact: "Promises earn this rating when they accomplish substantially less than the official's original statement but when there is still a significant accomplishment that is consistent with the goal of his original promise."

a training manual first developed for use by PolitiFact Texas in 2010. Here is what one journalism scholar says about PolitiFact's practices and procedures:

> Of the political fact-checking groups, PolitiFact has the most carefully articulated process for researching political claims and rendering a verdict. In part, this stems from the group's franchise model. In order to license the PolitiFact method to state partners, the group had to codify that method; it had to develop policies and procedures that could be imparted in training sessions and incorporated into official literature. (Graves 2016)

PolitiFact's policies and procedures include the following rules: PolitiFact's employees are required to avoid public expression of political opinion (including on social media) or have other public involvement in politics; they are not allowed to make political contributions; and they are not allowed to work on political campaigns. The process is to scan television, newspaper, and social media to identify political claims to fact-check. PolitiFact only assesses factual claims, not opinions. It limits itself to political claims that have been, are, or are likely to be circulated widely; and they are primarily interested not in esoteric topics but instead in topics likely to be of interest to the ordinary citizen. They try to be even-handed in the source of the claims they evaluate: "Without keeping count, we try to select facts to check from both Democrats and Republicans. At the same time, we more often fact-check the party that holds power or people who repeatedly make attention-getting or misleading statements."[16]

Other PolitiFact's processes try to reinforce this practice of care and even-handedness: PolitiFact publishes a list of sources for every fact-check. It provides links to sources. Its goal is to let people decide for themselves. In the investigation of a political claim or position, PolitiFact's staff typically conducts a Google search as well as a search of some online databases, consults experts, and reviews publications. This research emphasizes use of primary sources and original documentation. The research considers both government reports and academic studies. PolitiFact's editorial staff checks ratings and evidence of journalists. Stories are updated when new information is found or errors are discovered, but the original article is archived intact. In other words, facts are independently verified.

Table 5.3 provides a sample of PolitiFact's Truth-O-Meter in action – two judgments of each type that they made in mid-2018, one from a Democratic source, the other from a Republican source.

One might expect Conservatives to be critical of PolitiFact for its "bias". Indeed, there have been a number of Conservative criticisms, e.g. by the Republican National Committee in 2016 (Wolf 2016). Fuel was poured onto the fire when a University of Minnesota political science professor, Eric Ostermeier, published a study in 2011 showing that about an equal number of Republican and Democratic assertions were chosen for review by PolitiFact, but the assessments were three times as often false for the Republican claims (Ostermeier 2011). An independent

---

[16]The Principles of the Truth-O-Meter: PolitiFact's methodology for independent fact-checking, http://www.politifact.com/truth-o-meter/article/2018/feb/12/principles-truth-o-meter-politifacts-methodology-i/, accessed 27 June 2018).

Table 5.3 A sample of fact-checking from PolitiFact[a]

Who	What	When	Evaluation
Donald Trump	"Democrats mistakenly tweet 2014 pictures from Obama's term showing children from the Border in steel cages."	May 29, 2018	True
Bernie Sanders	"You know what Amazon paid in federal income taxes last year? Zero."	May 3, 2018	True
Donald Trump	"There's been a 1700 percent increase in asylum claims over the last 10 years."	June 21, 2018	Mostly True
Bernie Sanders	"We've spent more money on the military than the next 12 nations combined."	April 24, 2018	Mostly True
Mitch McConnell	"The bottom line is clear: Under the policies of this unified Republican government, American workers, families, and business owners are achieving economic growth that is unmatched in recent memory. ... More than 1 million new jobs have been created just since we passed tax reform last December."	June 5, 2018	Half True
Congresswoman Jacky Rosen (D-NV)	Dean Heller helped "craft a partisan repeal bill that also would have slashed coverage protections for people with pre-existing conditions."	June 13, 2018	Half True
Donald Trump	"We've already started (the border wall). We started it in San Diego."	June 26, 2018	Mostly False
Bill Clinton	"I left the White House $16 million in debt."	June 5, 2018	Mostly False
Donald Trump	"Crime in Germany is way up."	June 18, 2018	False
Nancy Pelosi	"Two-thirds of the people who use Medicaid are poor children, but two-thirds of the money is for long-term care for seniors, whether in a facility or at home."	May 11, 2018	False
Donald Trump	"The Failing @nytimes quotes 'a senior White House official,' who doesn't exist, as saying 'even if the meeting were reinstated, holding it on June 12 would be impossible, given the lack of time and the amount pf planning needed.' WRONG AGAIN! Use real people, not phony sources."	May 29, 2018	Pants on Fire!
LaWana Mayfield (D, Charlotte, NC City Council)	"I am still waiting for someone to produce pieces of the alleged plane that caused the Twin Towers to collapse."	April 25, 2018	Pants on Fire!

[a]Source: politifact.com

study in 2016 by political scientists Stephen Farnsworth and Robert Lichter reached similar findings to those of Ostermeier. They found "greater deceit in Republican rhetoric" when they analyzed PolitiFact. They also found that:

legislators who had more than one statement analyzed during the study period were dispro-
portionally likely to be influential members of the House or Senate leadership or likely
2016 presidential candidates. The lawmakers selected for greater scrutiny were also more
likely to be more ideologically extreme than the median members of their party caucuses.
(Farnsworth and Lichter 2016)

Another Conservative criticism came from Matt Shapiro, a writer for *The Federalist*. His simple data analysis of PolitiFact led to the following result:

When fact-checked by PolitiFact, Democrats had an average rating of 1.8, which is between
"Mostly True" and "Half True." The average Republican rating was 2.6, which is between
"Half-True" and "Mostly False." We also checked Republicans without President-elect
Donald Trump in the mix and found that 0.8 truth gap narrowed to 0.5. (Shapiro 2016)

Shapiro reports that, "PolitiFact rated Hillary Clinton and Tim Kaine as two of the most honest people among the 20 politicians we included in our data scrape (1.8 and 1.6, respectively) while Trump was rated as the most dishonest (3.2)" (Shapiro 2016).

Some Conservatives did not need the benefit of data analysis to criticize PolitiFact for political bias and questionable findings. Consider Kevin Williamson, a writer for *National Review*. Using strong, colorful language, he argued that PolitiFact had "intellectual dishonesty", "risibly sloppy and shockingly … lazy reporting habits", and a "feckless, gormless, and in any intelligent world unemployable [editor]" (Williamson 2015). He went on to talk about PolitiFact's double standards: "progressives mainly like to talk about science when it can be used as a cudgel for their moral program (regarding homosexuality, for example) or when it can be used to annoy or embarrass conservative Christians". He concluded: "deploying rank and obvious intellectual dishonesty in the service of narrow, partisan political sympathies. It is detestable, and it deserves to be condemned by all those who care about newspapers – not only by the conservatives against whom its intellectual dishonesty is directed" (Williamson 2015).

While one might expect political bias claims from Conservatives because that is what we see a great deal in today's press, it was the Liberals who were the first to question PolitiFact about political bias. The occasion was in 2011, when PolitiFact awarded the Lie of the Year to the Democratic Congressional Campaign Committee claim that the House had voted to "end Medicare". Various political organizations howled, including *Politico* and *Salon* (Hudson 2011; Marx 2012). In fact, the Democratic Party of Wisconsin, claiming persistent bias, decided in 2011 not to take calls from PolitiFact reporters (Gunn 2011).

At least four commentators complained about the process of political fact-checking, independent of any political bias. Ben Smith from *Politico* objected to the fact that "the framing implicitly exalts a certain class of "fact-finding" journalists above workaday hacks" (Marx 2012). Alec MacGillis of *The New Republic* argued that, "there are problems with assigning specific teams of reporters to call bullshit on political nonsense, rather than expecting all journalists to do so in the course of their work" (Marx 2012). Matt Welch criticized all fact-checking organizations because they focus on politicians' statements about their opponents and the opponents' policies, rather than on statements about their own policies (Welch 2013). James Poniewozik complains that Politifact's rating system hurts rather than helps political debate because it is so imprecise:

> [Spinning, shading the truth, insinuating, using logical fallacies, making dubious subjective interpretation] are as dangerous, if not more so, as blatant falsehoods of fact... So I get why PolitiFact needs to address this stuff... But the whole reason that PolitiFact exists is that *words and facts matter*. PolitiFact's job is important because inaccurate-but-catchy language, deployed a certain way and repeated, can create false impressions and misinform people. So it is with "Pants on Fire" here. To an average listener, PolitiFact is not saying that Reid has made a statement he can't prove. They're saying that he's made a statement that is actually false–and thus, they seem to certify that Romney has in fact paid taxes for years in which we don't know one way or another. (Poniewozik 2012)

Whatever concerns people might have about bias in political fact-checking or in the process of fact-checking itself, political fact-checking was here to stay as fake facts became a mainstay of politics as the candidates and parties geared up for the 2016 elections.

5.3   Run-Up to the 2016 Presidential Election[17]

It was clear even before the 2016 national elections that politics were becoming increasingly partisan and that fake news was increasingly becoming a problem. This section identifies four initiatives that began in the year or two prior to the presidential campaign: (1) The American Press Institute initiated a program for small newspapers and small media outlets to improve their political fact-checking in the new environment of the digital age. (2) The University of Missouri's journalism school initiated a project known as the Trusting News Initiative to teach journalists techniques to inspire greater public trust in the media. (3) The Data & Society Research Institute in New York City established a Media Manipulation Initiative, under the director of well-known Microsoft Research scholar danah boyd, to carry out interdisciplinary research and put it into action. (4) Santa Clara University's Markkula Center for Applied Ethics created Trust Indicators that unpack the ethics and the standards for fairness and accuracy in news organizations.

---

[17] Another organization founded in the lead up to the 2016 U.S. presidential elections is Polygraph. info. It is a nonpartisan fact-checking website, founded in 2015, run by Voice of America and Radio Free Europe/Radio Liberty. It is headquartered in Washington, DC. While it covers worldwide politics and has an international audience, it does cover some US politics, especially issues related to foreign policy, immigration, and human rights. One might question the objectivity of this organization, given that it is run by the U.S. government for propaganda purposes. However, Media Bias/Fact Check reports that Polygraph.info is "least biased" – leaning neither left nor right; and is rated VERY HIGH in its factual reporting. (https://mediabiasfactcheck.com/polygraph-info/, accessed 20 December 2018) We will not provide a more extensive profile of Polygraph.info here because it is mostly external-facing, addressing audiences outside the United States. It is an important international player, broadcasting its weekly programming in 45 languages to an audience upwards of 200 million people. (https://www.polygraph.info/p/5981.html)

The American Press Institute was established in 1946 as an educational nonprofit affiliated with the Newspaper Association of America. Today, its goal is to "help the news media, especially local publishers and newspaper media, advance in the digital age."[18] In 2014, it received a $400,000 grant from the Democracy Fund[19] for a 2-year project to improve political fact-checking, especially in small news organizations (American Press Institute 2014). The idea of this project, which has now become a permanent activity of the American Press Institute, known as its Accountability Journalism Program, was for the organization to "use its extensive networks within the news media, along with its credibility as a research group, to advance, refine, and defend this vital journalistic practice."[20] The work involved coordinating research of scholars at six universities during the first year, and holding workshops and producing resources to spread this knowledge of fact-checking during the second year of the grant, with a specific goal of preparing for the 2016 presidential election.

One of these resources on the website is a fact-checking primer. The primer points out that: "Fact-checking is a form of accountability journalism because the statements are typically made by people in politics, government or other powerful positions who are held accountable for their words and actions. Fact-checks require original sources, intense research, and high-quality data" (Elizabeth 2017). The primer emphasizes the importance of verification: "Some of that information is more reliable than others, and increasingly have no basis in reality at all. People need to be assured that professional media can help determine the reliability and accuracy of quotes, data and facts" (Elizabeth 2017). Political fact-checking, they claim their research shows, both makes politicians "less likely to misinform voters and misstate facts" and, for the newspapers, "increase[s] readership and audience engagement."[21] Interestingly, they express some doubts about the presentation methods used by others such as The Fact Checker of the *Washington Post*, with its Pinocchio rating system:

> In some newsrooms, fact-checking includes a ranking of the veracity of the statement being checked. Research shows that those rankings – which often are accompanied by colorful illustrations and catchy captions – does tend to attract more readers. Research also shows, however, that a rating system does not improve learning or comprehension among readers. Considering the current politically divisive climate in many communities, essentially calling statements "lies" or their speakers "liars" may not be the best way to promote

---

[18] https://www.americanpressinstitute.org/news-releases/american-press-institute-announces-major-project-improve-fact-checking-journalism/

[19] "The Democracy Fund invests in social entrepreneurs working to ensure that our political system is responsive to the priorities of the American public and has the capacity to meet the greatest challenges facing our country." ("American Press Institute announces major project to improve fact-checking journalism", news release, American Press Institute, February 6, 2014, https://www. americanpressinstitute.org/news-releases/american-press-institute-announces-major-project-improve-fact-checking-journalism/ (accessed 20 December 2018).)

[20] American Press Institute 2014. Also see https://www.americanpressinstitute.org/category/fact-checking-project/

[21] Elizabeth (2017). Their claims are based on American Press Institute's Metrics for News.

engagement. Newsrooms should weigh the sentiment among their audiences and consider creating attractive and engaging formats rather than ratings, and focusing on the fact-checking of *issues* rather that politicians' statements. (Elizabeth 2017)

Instead, the primer focuses on clarity, transparency, and original sourcing as the foundation of effective fact-checking. Inexperienced reporters need more training in research methods and in finding and evaluating sources, and even experienced reporters may need to improve the clarity and effectiveness of their writing. As they explain the task of fact-checking:

> Fact-checkers need to possess the basic characteristics of journalists – and then some. Curiosity, a desire to both learn and inform, and the ability to understand issues quickly are important. But in fact-checking, those issues can be exceptionally complicated and nuanced. Finding the right sources for verification can be exceptionally difficult. Presenting them to people with clarity and context can be an exercise in creativity and even psychology. (Elizabeth 2017)

The American Press Institute wants individual journalists to learn more about the concepts and strategies behind fact-checking, so they have provided a series of educational articles on their website. These include: a guide to verification and debunking words and images; a description of how experts define the concepts of fake news, misinformation, and disinformation; how to evaluate a source on the CRAP test (currency, reliability, authority, and purpose, which was developed by communication scholar Howard Reingold[22]); and a guide to about 200 fake news sites. This site also provides 11 articles providing tactics for journalists and news institutions who want to do a better job fact-checking. Some examples include: discussion of primary versus secondary sources; guidelines for newsrooms new to fact-checking; how to identify misleading political talking points; and best practices from some other organizations such as FactCheck.org and PolitiFact.[23]

In early 2016, the University of Missouri journalism school, one of the oldest and most distinguished in the country, established the Trusting News initiative.[24] Public trust in journalism had been sinking, and, according to a Gallup Poll in 2016 only 32% of Americans indicated "a great deal" or "a fair amount" of trust in mass media.[25] The idea was to teach journalists techniques that would enable them to gain the trust of their potential audience, and convince people to use them as the go-to source for news and to spread their news stories rather than fake news stories on social media. The overall idea was to "reclaim the credibility of journalism."[26]

---

[22] See, for example, Reingold (2013).

[23] https://betternews.org/topic/fact-checking/ (accessed 20 December 2018).

[24] In 2018, the University of Missouri journalism school took in the University of Georgia journalism school as a partner on the Trusting News initiative. ("Trusting News Project Expands Research and Training Through Partnership with University of Georgia," *RJI Online*, October 11, 2018, https://www.rjionline.org/stories/trusting-news-project-expands-research-and-training-through-university-of-g

[25] Gallup Poll as reported in Mayer (2017).

[26] https://www.rjionline.org/stories/series/trusting-news?utm_source=newsletter&utm_medium=email&utm_campaign=newsletter_axiosam&stream=top, accessed 3 January 2019.

When Joy Mayer, a professor in the journalism school, began the project, it was not yet clear the methods or events that would arise in the 2016 presidential election that would make their project even more urgent:

> When we began the Trusting News project in January 2016, we had no idea how the presidential campaign would evolve. We didn't know the intentional spread of false information would play an even larger role in the information climate. We didn't know Facebook's algorithm would move toward favoring posts shared by individuals over those shared by pages, making it all the more important that news consumers help spread our content.[27]

The project was launched with a $100,000 grant from the Knight Foundation.[28] The project involved questionnaires, journalist logs, interviews with hundreds of journalists in over 50 newsrooms, and building a toolkit that could be used by journalists to build trust with the public. Some of the newspapers involved include the *Cincinnati Enquirer, Dallas Morning News, Minneapolis Star Tribune, Fresno Bee, Fort Worth Star-Telegram* and the *Ogden* (Utah) *Standard-Examiner* (Stroh 2017). The project conducted research to determine what leads the public to trust journalists:

> We all know trust builds over time. That also applies in the relationship between journalists and their communities. Trust in a news organization develops when people know they can turn to you consistently for reliable information. It happens when people feel they are being heard. It happens when they see their own lives and priorities reflected in your news coverage. It happens when they have confidence in the decisions, values and ethics taking place in your newsrooms. (Mayer and Walsh 2018)

Of course, the big question was how journalists were to put this into action:

> But if you do not tell them about the editorial decisions you are making on a daily basis or listen and respond to their feedback (even the critical comments), how do they know they are being heard or that you are thinking critically about your news coverage?
>
> At the Trusting News project we are learning that one key area of disconnect between the public and journalists is that news consumers often don't realize what journalism encompasses. When they hear the word "journalism," they're likely not thinking it is the story that informed them of how many new teachers their child's school is hiring or what the city's plans are for the beach that's eroding or what new businesses are opening at the mall. And partly because we do not consistently tell them how we do our jobs, they assume the worst about our motivations, ethics and commitment to accuracy.
>
> They assume we quote people without fact-checking what they tell us. They assume we pay people to speak about certain topics. They assume we publish all information without considering the effect it could have on people....[29]

Table 5.4 presents some of the major findings of the project.

---

[27] https://www.rjionline.org/stories/series/trusting-news?utm_source=newsletter&utm_medium=email&utm_campaign=newsletter_axiosam&stream=top, accessed 3 January 2019.

[28] "Trusting News Project Receives $100,000 from Knight Foundation" *RJI Online*, September 25, 2017, https://www.rjionline.org/stories/trusting-news-project-receives-100000-grant-from-knight-foundation. Other funding came from the Democracy Fund and from the university's journalism institute.

[29] Mayer and Walsh, August 16, 2018.

Table 5.4 Some major finds from the trusting news initiative[a]

Tell Your Story
Demonstrate that your newsroom reflects your audience's values.
Look for reasons to point out what makes your staff credible or knowledgeable on a specific topic, but focus on the topic more than the staff.
Explain the process behind a high-interest newsroom decision or project.
Engage Authentically
Reward productive comments and publicly challenge harmful ones.
Look for ways to inject your own emotion or perspective into your writing.
Especially as you get started, focus on hosting and participating in conversations people are eager to have. As you earn your users' participation, they'll be being willing to jump in on more nuanced or difficult topics.
Be a vehicle for users to share strong emotions – especially appreciation, outrage, pride and nostalgia.
Try making conversation the purpose of a post (rather than treating it as subordinate to the sharing of a link).
Deploy Your Fans
Encourage sharing on posts and topics users will WANT to share, not ones you WISH they would share.
Start with easy asks (posts that are obviously helpful, positive or fun).
Invite sharing on posts that are in the public interest. When police need help finding a missing child, or a food recall endangers peoples' health, lots of users will see it as a civic duty to help spread the word.
Emphasize emotion when appropriate. If users are likely to share because they're proud, outraged, scared or excited, consider whether the framing of your post includes that angle.

[a]Source: https://trustingnews.org/strategy-results/

In 2013, danah boyd, the well-known senior researcher at Microsoft Research, came up with the idea for a research institute "focused on social, cultural, and ethical issues arising from data-centric technological development."[30] The Data & Society Research Institute was established in 2014 with a grant from Microsoft, and boyd serves as the chair of the board of directors. Since then, numerous major foundations (Sloan, MacArthur, Gates, Ford, Knight, Newmark, Robert Wood Johnson, Kellogg, Solidago, Omidyar) have provided general or project support[31] The research institute "is committed to identifying issues at the intersection of technology and society, providing research that can ground public debates, and building a network of researchers and practitioners who can offer insight and direction."[32] The cross-cutting themes for their research agenda are: vulnerabilities in socio-technical systems; systems of rights, equity, and governance in a networked world; and intersections of humans and intelligent systems.[33]

---

[30] danah boyd, letter from the president, https://datasociety.net/pubs/ar/DS_Report-on-Activities_2014-2015.pdf

[31] https://datasociety.net/about/#recent_press

[32] 2014–15 annual report, https://datasociety.net/pubs/ar/DS_Report-on-Activities_2014-2015.pdf

[33] https://datasociety.net/research/

In 2016, the Media Manipulation Initiative was created by Data & Society. According to its web page, this initiative

> works to provide news organizations, civil society, platforms, and policymakers with insights into new forms of media manipulation to ensure a close and informed relationship between technical research and socio-political outcomes. This requires assessing strategic manipulation, imagining the possibilities for encoding fairness and accountability into technical systems, and conducting ethnographic research to describe and understand new social activity.[34]

The researchers use a sociotechnical approach to understand "the social, political, and economic incentives to game information systems, websites, platforms, and search engines – especially in cases where the attackers intend to destabilize democratic, social, and economic institutions."[35] They study various kinds of manipulation practices:

> planting and/or amplifying misinformation and disinformation using humans (troll armies, doxxing, and bounties) or digital tools (bots); targeting journalists or public figures for social engineering (psychological manipulation); gaming trending and ranking algorithms, and coordinating action across multiple user accounts to force topics, keywords, or questions into the public conversation.[36]

Topics include analyzing the use of digital infrastructures for political purposes; the tools these groups use; audience engagement with these efforts; and the economic, legal, and regulatory factors at play.

The research effort is led by sociologist and science studies scholar Joan Donovan. The interdisciplinary research team includes media historians and theorists, communication scholars, media and information policy scholars, a digital media folklorist, and computer and data scientists. One of the main outputs of this initiative are reports on such topics as a lexicon of terms used for lying and misrepresentation (Jack 2017), how hate groups took advantage of the news media (Phillips 2018), an ethnographic study of media practices in conservative Christian communities (Tripodi 2018), and potential approaches to containing fake news (trust and verification, disrupting economic incentives, de-prioritizing content and banning accounts, and regulation) (Caplan et al. 2018).

Another initiative comes from the other side of the nation, at the Markkula Center for Applied Ethics at Santa Clara University in the heart of Silicon Valley. Fake news is only one issue of concern to this center; they also are interested, for example, in homelessness and immigration issues. The project, which began in 2015 but picked up steam after the 2016 election, is led by Sally Lehrman, a Peabody Award-winning journalist who heads the journalism ethics program at the Center.[37] The goal is to produce transparency standards in collaboration with more than 70 leading news

---

[34] https://datasociety.net/research/

[35] https://datasociety.net/research/media-manipulation/?utm_source=newsletter&utm_medium=email&utm_campaign=newsletter_axiosam&stream=top (accessed 191,012).

[36] Ibid.

[37] "Sally Lehrman", Markulla Center for Applied Ethics, Santa Clara University, https://www.scu.edu/ethics/about-the-center/people/sally-lehrman/

Table 5.5   The initial trust indicators from the Markulla Center Trust Project[a]

Best Practices: What are the news outlet's standards? Who funds it? What is the outlet's mission? Plus commitments to ethics, diverse voices, accuracy, making corrections and other standards.
Author/Reporter Expertise: Who made this? Details about the journalist, including their expertise and other stories they have worked on.
Type of Work: What is this? Labels to distinguish opinion, analysis and advertiser (or sponsored) content from news reports.
Citations and References: What's the source? For investigative or in-depth stories, access to the sources behind the facts and assertions.
Methods: How was it built? Also for in-depth stories, information about why reporters chose to pursue a story and how they went about the process.
Locally Sourced? Was the reporting done on the scene, with deep knowledge about the local situation or community? Lets you know when the story has local origin or expertise.
Diverse Voices: What are the newsroom's efforts and commitments to bringing in diverse perspectives? Readers noticed when certain voices, ethnicities, or political persuasions were missing.
Actionable Feedback: Can we participate? A newsroom's efforts to engage the public's help in setting coverage priorities, contributing to the reporting process, ensuring accuracy and other areas. Readers want to participate and provide feedback that might alter or expand a story.

[a]"Frequently Asked Questions," The Trust Project, https://thetrustproject.org/faq/#indicator

organizations in Europe and the United States (including *The Economist*, *The Globe and Mail*, Hearst Television, *La Repubblica*, and the *Washington Post*) and search engine and social media companies (Google, Facebook, Bing). After interviews in both the United States and Europe, they learned: "It turns out they don't just want to weed out imposters. They want to know who wrote or produced a story, what expertise they have, and whether the publisher has an agenda. Transparency matters."[38] Thus, Lehrman created a digital tool of these Trust Indicators in collaboration with these news organizations and the search engine and social media companies. Trust Indicators are "standardize[d] disclosures that provide clarity on a news organization's ethics and other standards for fairness and accuracy, a journalist's background, and the work behind a news story."[39] The eight Trust Indicators that the project decided to implement first are presented in Table 5.5.[40]

---

[38] "Who We Are," The Trust Project, https://thetrustproject.org/?utm_source=newsletter&utm_medium=email&utm_campaign=newsletter_axiosam&stream=top, accessed 3 January 2019).
[39] "What are the Trust Indicators?" https://thetrustproject.org/?utm_source=newsletter&utm_medium=email&utm_campaign=newsletter_axiosam&stream=top
[40] For a discussion of the Trust Indicators and how they came about, see Lehrman (2017).

5.4   Following the 2016 Presidential Election[41]

Following the 2016 election and the first few months of the Trump Administration, numerous new initiatives were started to identify fake facts and improve the quality of journalism. There have been two notable patterns in these recent efforts: the use of technology to augment or replace human evaluators, and the rise of for-profit operations. We examine seven of these recent initiatives in this section: (1) *The Weekly Standard* Fact Check is like traditional fact-checking operations such as those of the *Washington Post* or PolitiFact, but it differs by being carried out by a Conservative magazine. (2) The CUNY journalism school has established the News Integrity Initiative, which is largely a news literacy project, which conducts research and funds projects of other organizations. (3) The Duke University Reporter's Lab has established the Tech & Check Cooperative, bringing journalists and computer scientists together to develop automated fact-checking tools for journalists to use. (4) NewsGuard, created by well-known journalists Steven Brill and Gordon Crovitz, uses "nutrition labels" to characterize the news quality of individual online news sites. (5) The for-profit organization our.news has developed a crowdsourcing platform to fact-check and rate news. (6) Jimmy Wales, the co-founder of Wikipedia, has created a wiki-style, for-profit organization called WikiTribune that blends the use of professionals and volunteers to crowdsource the creation of news stories. (7) For-profit Deepnews.ai uses artificial intelligence (neural nets) to identify quality news stories online. In many of these cases, it is too soon to evaluate their success. However, WikiTribune has had to fire all of its staff in 2018 for lack of funding, so it continues as a volunteer-only activity; and *The Weekly Standard* ceased publication in December 2018.

Some of the fact-checking organizations, such as those of the *Washington Post* and PolitiFact, are seen as left-leaning – even if we saw evidence earlier in this chapter that disputes their bias. One approach that appeared after the 2016 election was for a right-leaning publication, *The Weekly Standard*, to create its own fact-checking operation, known as TWS Fact Check. *The Weekly Standard* runs a fact-checking section on its website.[42] *The Weekly Standard*, which was created by William Kristol and Fred Barnes in 1995, only started its fact-checking page in 2017. Its fact-checker is Holmes Lybrand, who used to work for Transparency Texas, a nonprofit organization focused on campaign finance. The staff searches for claims in the media that are not obviously supported by evidence and accepts requests from the public to investigate various political claims in the news. Like

---

[41] There are similar initiatives since 2016 that are directed primarily outside the United States, but we will not examine them in this chapter. One example is the Journalism Trust Initiative sponsored by Reporters Without Borders, and Agence France Presse, the European Broadcasting Union and the Global Editors Network. See, for example, "RSF and its partners unveil the Journalism Trust Initiative to combat Disinformation," *Reporters Without Borders*, April 3, 2018, https://rsf.org/en/news/rsf-and-its-partners-unveil-journalism-trust-initiative-combat-disinformation (accessed 3 January 2019).

[42] https://www.weeklystandard.com/tag/fact-check

other fact-checking organizations Fact Check does research in documents and consults subject-area experts, then provides citations and direct links to online sources as a means to exhibit the evidence behind their evaluation. TWS Fact Check does not have a rating system, such as Pinocchios to be awarded. "Further, the TWS Fact Checker is prohibited from writing opinion pieces of the sort that would appear elsewhere on The Weekly Standard and is explicitly directed to avoid engaging in political parties or advocacy organizations."[43]

In 2017, the City University of New York's Newmark Graduate School of Journalism created the News Integrity Initiative. It is a $14 million fund to support "efforts to connect journalists, technologists, academic institutions, non-profits, and other organizations from around the world to foster informed and engaged communities, combat media manipulation, and support inclusive, constructive, and respectful civic discourse."[44] The Initiative provides grants, conducts research (sometimes with partners) in the School, and holds events. They regard this effort as a kind of news literacy project. The Initiative is managed by Meredith Craven, a senior executive at Facebook for strategic partner development – integrity; Craig Newmark, the founder of Craigslist and major donor to the School; and Jeff Jarvis, the director of the Tow-Knight Center for Entrepreneurial Journalism (a center within the School).[45] Founders of the fund include Facebook, the Craig Newmark Philanthropic Fund, AppNexus, the Knight Foundation, the Ford Foundation, the Democracy Fund, the Tow Foundation, Mozilla, and Betaworks.[46]

Projects that the Initiative funds include ones to enhance transparency and cooperation between newsrooms and the public; ways to improve community conversations; and means to amplify marginalized voices, cultivate diversity, and diminish the influence of misinformation.[47] The scope is global. One project that is being partly funded currently by the Initiative is the Media Manipulation Initiative of Data & Society (described above). Another project the Initiative is funding is being carried out by the University of Florida journalism school on "what research from

---

[43] Fact Check, *The Weekly Standard*, https://www.weeklystandard.com/fact-check (accessed 4 January 2019). Media Bias/Fact Check gave *The Weekly Standard* (not its fact-checking operation) the overall rating of Right Bias, and stated: "These media sources are moderately to strongly biased toward conservative causes through story selection and/or political affiliation. They may utilize strong loaded words (wording that attempts to influence an audience by using appeal to emotion or stereotypes), publish misleading reports and omit reporting of information that may damage conservative causes. Some sources in this category may be untrustworthy." (https://mediabiasfactcheck.com/weekly-standard/, accessed 4 January 2019).

[44] News Integrity Initiative, https://www.journalism.cuny.edu/centers/tow-knight-center-entrepreneurial-journalism/news-integrity-initiative/ (accessed 3 January 2019).

[45] Advisory Council, https://www.journalism.cuny.edu/centers/tow-knight-center-entrepreneurial-journalism/news-integrity-initiative/advisory-council/ (accessed 3 January 2019).

[46] Founders, https://www.journalism.cuny.edu/centers/tow-knight-center-entrepreneurial-journalism/news-integrity-initiative/#founders (accessed 3 January 2017).

[47] What We Fund, https://www.journalism.cuny.edu/centers/tow-knight-center-entrepreneurial-journalism/news-integrity-initiative/what-we-fund/ (accessed 3 January 2019).

multiple academic disciplines tells us about community engagement and trust in news" and to implement best practices gleaned from that research (Ortiz 2018).

The Knight Foundation, Facebook, and the Craig Newmark Foundation came together in 2017 to provide $1.2 million to automate fact-checking in a project called the Tech & Check Cooperative, which is being carried out by the Duke University Reporter's Lab.[48] The project involves journalism and computer science faculty from Duke, the University of Texas at Arlington, and Cal Poly-San Luis Obispo to build tools and apps that can be used by journalists to automate such fact-checking tasks as mining information from transcripts, media streams, and social feeds. One product of this Cooperative is the FactStream iPhone app, which provides fact-checks in real time from FactCheck.org, PolitiFact, and the Washington Post. It was successfully tested on Donald Trump's State of the Union address in January 2018 (Adair 2018). The Cooperative also held a Tech & Check Conference at Duke in March 2018, which brought together 40 people from around the world interested in fact-checking. The conference gave technologists a chance to demo fact-checking technologies, such as Chequeabot, which provides an automated fact-checking service to the Argentinian fact-checker Chequeado, and the Bad Ideas Factory's Truth Goggles tool (Iannucci 2018).

Another approach, a for-profit venture, has been taken by two journalists Steven Brill (founder of American Lawyer, Court TV, and *Brill's Content* magazine) and Gordon Crovitz (*Wall Street Journal* publisher and columnist). Called NewsGuard, it was started in early 2018 with funding from the Knight Foundation and venture capital funds from the Publicis Groupe. The basic idea is to use teams of independent journalists to rate 7500 sources of online news for credibility and transparency. These sources, they claim, constitute the source of 98% of the news seen by Americans. NewsGuard gives each news website an overall green or red rating for whether they follow "basic standards of accuracy and accountability."[49] The team also writes up what the founders call "nutrition labels" on each website they rate, discussing "the site's background, ownership, content, and why it received its rating."[50] They have also created a browser plug-in that works with Bing, Facebook, Google, and Twitter that will make NewsGuard's rating icon appear each time someone calls up a news website.[51] It is hoped that the nutrition labels will be attractive to marketers and agencies, who desire some guarantee that they are working with legitimate news sites. Table 5.6 shows the content of a typical nutrition label.

Another example of a for-profit organization trying to solve the fake news problem is our.news, which was created by technologist Richard Zack in 2017. The

---

[48] Knight Foundation, Facebook and Craig Newmark provide funding to launch Duke Tech & Check Cooperative," September 25, 2017, https://reporterslab.org/duke-tech-and-check-cooperative-funding-announcement/?utm_source=newsletter&utm_medium=email&utm_campaign=newsletter_axiosam&stream=top (accessed 3 January 2019).

[49] "How It Works", NewsGuard, https://www.newsguardtech.com/how-it-works/ (accessed 3 January 2019).

[50] "How It Works".

[51] See Farhi (2018).

Table 5.6   Content provided on a NewsGuard nutrition label[a]

Overall rating – either a green check mark or a red X as to whether the site maintains basic standards of accuracy and accountability.
Credibility – either a green check mark or a red X for each of the following five criteria:
Does not repeatedly publish false content
Gathers and presents information responsibly
Regularly corrects or clarifies errors
Handles the difference between news and opinion responsibly
Avoids deceptive headlines
Transparency – either a green check mark or a red X for each of the following four criteria:
Website discloses ownership and financing
Clearly labels advertising
Reveals who's in charge, including any possible conflicts of interest
Provides information about content creators

[a]Source: "NewsGuard's trust initiative shows early promise," *Local Media Insider*, 12 December 2018, http://localmediainsider.com/stories/newsguards-trust-initiative-shows-early-promise,1639 (accessed 3 January 2019)

company has built an online crowdsourcing platform to fact-check and rate the news. The company provides various tools for members of the crowd to participate: browser extensions, mobile apps, websites, and platform integrations. They also provide a Newstrition Label (somewhat like NewsGuard's Nutrition Label, created jointly by our.news and the Freedom Forum Institute, which provides verified information about the news publishers). The website creates competition to participate by providing lists of who are the most active members of the crowdsourcing community doing these news ratings. The platform does not provide individual ratings, only aggregate numbers in the rating of each news source. The site is still in beta test; when formally released, it is anticipated it will use open-source software.[52]

Another for-profit attempt was WikiTribune, created in 2017 by Jimmy Wales, the co-founder of Wikipedia. Wales was stimulated to start it by a comment from Trump advisor Kellyanne Conway at the very beginning of Trump term in office, when she defended the false statement by White House Press Secretary Sean Spicer about the size of the Trump inauguration crowd, calling what Spicer said "alternative facts" (Hern 2017). Like Wikipedia, this was to be a crowd-sourced effort; but there was no connection between Wikipedia and WikiTribune. The original idea for WikiTribune was that individual volunteers would proofread, fact-check, and add additional sources to news articles, while a professional staff of journalists would conduct research, report news, and check the work of volunteers. As Wales remarked: "This will be the first time that professional journalists and citizen journalists will work side-by-side as equals writing stories as they happen, editing them live as they develop and at all times backed by a community checking and rechecking all facts" (Whigham 2017). It was intended to be like other sites in the wiki community, self-

---

[52] FAQ, our.news, updates September 19, 2018, https://our.news/faq/ (accessed 3 January 2018).

regulated, "creating news by the people and for the people."[53] Content was going to be made available under a Creative Commons license.

The organization was to be crowdfunded, beginning in April 2017. Enough funds were donated that the organization could initiate operations in October 2017, but the continuing level of donation was insufficient to pay the salaries of the ten professional staff members, and they were all laid off 1 year later. The idea of donors making monthly payments and suggesting topics for news coverage did not prove to be financially viable. Today, individuals can continue to do their volunteer work for WikiTribune as before, but there is no longer any professional review of materials.[54]

Several of the recently created organizations that we have discussed in this section have used technology as a means to fight fake facts. Frederic Filloux started a project using the technique from artificial intelligence known as *deep learning* to identify quality journalism online. Filloux had been a journalist for Groupe Les Echos, France's leading business news organization, until he left in 2015 to form *Monday Note*, a popular newsletter and blog that discusses the role of digital technology in transforming the media industries. In academic year 2016–17, he held a John S. Knight fellowship in the communication department at Stanford University.[55] During that time, he developed what was called the News Quality Scoring Project, to use automatic means to identify the quality of news stories. At the end of this year, he received a Knight Prototype Fund grant to commercialize this technology, and it is now called Deepnews.ai (Filloux 2017).

Individual stories are evaluated by the Deepnews engine on a 1–5 scale: "we look for value-added journalism. This means coverage built on a genuine journalistic approach: depth of reporting, expertise, investigation, analysis, ethical processes, and resources deployed by the newsroom."[56] The system uses neural nets to develop the scoring system.[57] Filloux explains why he uses artificial intelligence instead of

---

[53] Wales, quoted in Whigham (2017).

[54] On the various reasons why WikiTribune failed, see Ingram (2018); Lafrance (2017); Bell (2017).

[55] https://jsk.stanford.edu/fellows/class-of-2017/frederic-filloux/ (accessed 4 January 2019).

[56] https://www.deepnews.ai/?utm_source=newsletter&utm_medium=email&utm_campaign=newsletter_axiosam&stream=top (accessed 4 January 2018).

[57] "The platform is based on a combination of two models.

"– The first model involves two sets of "signals" to assess the quality of journalistic work: Quantifiable Signals and Subjective Signals. Quantifiable Signals are collected automatically. These signals include the structure and patterns of the HTML page, advertising density, use of visual elements, bylines, word count, readability of the text, information density (number of quotes and named entities). Subjective Signals are based on criteria used by editors (and intuitively by readers) to assess the quality of a story: writing style, thoroughness, balance & fairness, timeliness, etc. (This set will be used only in the building phase of the model).

"– The second model is based on deep learning techniques, like "text-embedding" in which texts from large volumes of data (millions of articles) are converted into numerical values to be fed into a neural network. This neural net returns probabilities of scoring." (https://www.deepnews.ai/?utm_source=newsletter&utm_medium=email&utm_campaign=newsletter_axiosam&stream=top, accessed 4 January 2019).

humans to score individual news articles (although humans are involved in training
the algorithms of the neural net):

> Humans don't scale. News aggregators process about 100 million new pieces of information
> per day, half of them in English. Artificial intelligence is the only way to process such a
> large stream. In the "Signals" model, none of the indicators, considered separately, say
> much about the quality of a story; only their combination does. Assigning the proper weight
> to each signal can't be done in a deterministic way, but it is a perfect job for a neural
> network. This is why the model is based on A.I.[58]

Filloux sees four benefits of Deepnews.ai. If the quality of journalism improves and
people know they are reading on a quality news site, they are more likely to read
more stories (currently, people read fewer than two stories, on average, on a given
news site). The system can customize the set of news articles selected for an
individual by knowing the reader's preferences and searching on the metadata of
individual stories. Once the quality of individual news stories can be determined,
advertising to be attached to the story can be priced accordingly, with higher quality
stories fetching higher advertising prices. Finally, Deepnews.ai can make news
publishers less reliant on companies such as Google and Apple in bringing readers
to material they are interested in.[59]

Filloux explicitly addressed the question of how Deepnews.ai relates to fake
news:

> Does Deepnews.ai address issues such as accuracy and fake news? Indirectly, yes. From
> what we have observed, fake news has a different structure from legitimate journalism. We
> are certain that it can be measured; for instance, we found some distinctive patterns in the
> sentence structure from different news sources. Also, by relying on the reputation of
> publishers and authors, and by taking into account their performance over time, Deepnews.
> ai should be able to downgrade suspicious content items.[60]

The business model for Deepnews.ai is to get news publishers, news distributors,
and advertising agencies to pay to access the Deepnews platform, with pricing
based on volume of articles.

5.5 Conclusions

We have seen that political fact-checking is something that began early in the
twenty-first century but has become increasingly common as the century has gone
on and as the use of fake facts as a political tool has increased. Political fact-checking
was stimulated in particular by the 2008 and 2016 national elections. We see a

---

[58] https://www.deepnews.ai/?utm_source=newsletter&utm_medium=email&utm_
campaign=newsletter_axiosam&stream=top (accessed 4 January 2019).

[59] https://www.deepnews.ai/?utm_source=newsletter&utm_medium=email&utm_
campaign=newsletter_axiosam&stream=top (accessed 4 January 2018).

[60] https://www.deepnews.ai/?utm_source=newsletter&utm_medium=email&utm_
campaign=newsletter_axiosam&stream=top (accessed 4 January 2019).

number of trends emerging: political fact-checking is becoming an international and not just a US activity. What was at first completely a human activity is becoming either supplemented by tools (such as apps that will check databases about the veracity of a claim) or supplanted by tools (such as the Deepnews.ai site making largely-human-independent decisions about the quality of online journalism). We are also seeing political fact-checking not as something that one needs to pay for to maintain the integrity of journalism and politics, but as an emerging business opportunity. Several of the most recent efforts have been made by for-profit organizations. What the correct revenue models are for these business ventures is not yet settled.

The lack of evidence of viable business models suggests a pattern of behavior evident, for example, in the 1960s with the arrival of small software companies, personal computer vendors in the 1970s, and app and smart phone vendors in the 2010s. As a need becomes evident – as here with the need for fact checking – people experiment until they find something that works and is economically sustainable. As of this writing in 2019, that does not seem to be the case with fact checkers, although there are many of them. What we can conclude based on the evidence presented in this chapter is there is a clear need for fact checking for more than just for journalists or academics – the public at large has the same need.

References

Adair, Bill. 2018. *FactStream App Now Shows Latest Fact-Checks from Post, FactCheck.org, and PolitiFact*, October 7. https://reporterslab.org/factstream-app-now-shows-latest-fact-checks-from-post-factcheck-org-and-politifact/. Accessed 3 Jan 2018.

Arendt, Hannah. 1971. Lying in Politics: Reflections on the Pentagon Papers. *New York Review of Books,* November 18. http://www.nybooks.com/articles/1971/11/18/lying-in-politics-reflections-on-the-pentagon-pape/. Accessed 11 July 2018.

Bell, Emily. 2017. Wikitribune Venture Will Not Address Journalism's Underlying Issues. *The Guardian*, April 30. https://www.theguardian.com/media/2017/apr/30/wiktribune-experiment-will-not-address-journalisms-underlying-issues. Accessed 3 Jan 2019.

Caplan, Robyn, Lauren Hanson, and Joan Donovan. 2018. *Dead Reckoning: Navigating Content Moderation After Fake News*, February 21. https://datasociety.net/output/dead-reckoning/.

Dewey, Caitlin. 2016. Facebook Has Repeatedly Trended Fake News Since Firing Its Human Editors. *Washington Post,* October 12. https://www.washingtonpost.com/news/the-intersect/wp/2016/10/12/facebook-has-repeatedly-trended-fake-news-since-firing-its-human-editors/?noredirect=on&utm_term=.9e5ac5b05c51. Accessed 4 Jan 2019.

Elizabeth, Jane. 2017. Fact-Checking: A primer of the American Press Institute. *Better News*, September. https://betternews.org/fact-checking-primer/. Accessed 20 Dec 2018.

Farhi, Paul. 2018. A Journalistic Fix for Fake News? A New Venture Seeks to Take on the Epidemic. *The Washington Post*, March 3. https://www.washingtonpost.com/lifestyle/style/a-journalistic-fix-for-fake-news-a-new-venture-seeks-to-take-on-the-epidemic/2018/03/02/065438ca-1daf-11e8-b2d9-08e748f892c0_story.html?noredirect=on&utm_term=.115a5cc347b1. Accessed 3 Jan 2019.

Farnsworth, Stephen J., and S. Robert Lichter. 2016. *A Comparative Analysis of the Partisan Targets of Media Fact-Checking: Examining President Obama and the 113th Congress*. Paper Delivered at the American Political Science Association Annual Meeting, Philadelphia,

September. https://www.americanpressinstitute.org/wp-content/uploads/2016/10/2016-apsa-politifact.pdf. Accessed 27 June 2018.

Filloux, Frederic. 2017. The News Quality Scoring Project: Surfacing Great Journalism from the Web. *Monday Note*, June 25. https://mondaynote.com/the-news-quality-scoring-project-surfacing-great-journalism-from-the-web-48401ded8b53. Accessed 4 Jan 2018.

Graves, Lucas. 2016. *Deciding What's True: The Rise of Political Fact-Checking in American Journalism.* New York: Columbia University Press.

Gunn, Eric. 2011. No Comment. *Milwaukee Magazine,* June 14. https://www.milwaukeemag.com/NoComment/. Accessed 27 June 2018.

Hern, Alex. 2017. Wikipedia Founder to Fight Fake Newswith New Wikitribune Site. *The Guardian,* April 24. https://www.theguardian.com/technology/2017/apr/25/wikipedia-founder-jimmy-wales-to-fight-fake-news-with-new-wikitribune-site. Accessed 3 Jan 2019.

Hudson, John. 2011. Politifact Just Lost the Left. *The Atlantic,* December 20. https://www.the-atlantic.com/politics/archive/2011/12/politifact-just-lost-left/334047/. Accessed 27 June 2018.

Iannucci, Rebeca. 2018. *Journalist, Computer Scientists Gather for Tech & Check Conference at Duke,* March 30. https://reporterslab.org/tech-and-check-conference-fact-checking-duke-university/. Accessed 3 Jan 2019.

Ingram, Mathew. 2018. Wikipedia's Co-founder Wanted to Let Readers Edit the News. What Went Wrong? *Columbia Journalism Review,* November 19. https://www.cjr.org/analysis/jimmy-wales-wikitribune.php. Accessed 3 Jan 2019.

Jack, Caroline. 2017. *Lexicon of Lies,* August 9. https://datasociety.net/output/lexicon-of-lies/.

Kessler, Glenn. 2013a. About the Fact Checker. *Washington Post,* September 11. https://www.washingtonpost.com/news/fact-checker/about-the-fact-checker/?utm_term=.e25b71056798. Accessed 20 Dec 2018.

———. 2013b. *About the Fact Checker,* September 11. https://www.washingtonpost.com/news/fact-checker/about-the-fact-checker/?utm_term=.5fb05cc8b9a3. Accessed 4 Jan 2019.

Lafrance, Adrienne. 2017. The Problem with WikiTribune. *The Atlantic,* April 25. https://www.theatlantic.com/technology/archive/2017/04/wikipedia-the-newspaper/524211/. Accessed 3 Jan 2019.

Lehrman, Sally. 2017. What People Really Want from News Organizations. *The Atlantic,* May 25. https://www.theatlantic.com/technology/archive/2017/05/what-people-really-want-from-news-organizations/526902/. Accessed 3 Jan 2019.

Marx, Greg. 2012. What the Fact-Checkers Got Wrong: The Language of Politifact and Its Peers Does Not Match Their Project. *Columbia Journalism Review* (January 5). https://archives.cjr.org/campaign_desk/what_the_fact-checkers_get_wro.php. Accessed 27 June 2018.

Mayer, Joy. 2017. Journalists, Let's Invest in Trust, Not Just Expect It. *Trusting News*, November 17. https://www.rjionline.org/stories/journalists-lets-invest-in-trust-not-just-expect-it. Accessed 3 Jan 2019.

Mayer, Joy, and Lynn Walsh. 2018. Journalists: Defend Your Work Through Action, Not Just with Editorials. *Trusting News*, August 16. https://www.rjionline.org/stories/journalists-defend-your-work-through-action-not-just-with-editorials.

Minor, Jordan. 2016. *Facebook Floods the Internet with Lies, Google Tries to Bring Back Truth,* October 14. https://www.geek.com/tech/facebook-floods-the-internet-with-lies-google-tries-to-bring-back-truth-1675103/. Accessed 4 Jan 2019.

Ortiz, Rosaleen. 2018. *New Integrity Initiative and University of Florida to Help Newsrooms Increase Media Trust,* April 30. https://www.journalism.cuny.edu/2018/04/uf-news-integrity-initiative-partner-assist-newsrooms-use-cognitive-science-help-increase-media-trust/. Accessed 3 Jan 2019.

Ostermeier, Eric. 2011. Selection Bias? PolitiFact Rates Republican Statements as False at 3 Times the Rate of Democrats. *Smart Politics*, February 10. http://editions.lib.umn.edu/smartpolitics/2011/02/10/selection-bias-politifact-rate/. Accessed 27 June 2018.

Phillips, Whitney. 2018. *The Oxygen of Amplification: Better Practices for Reporting on Extremists, Antagonists, and Manipulators,* May 22. https://datasociety.net/output/oxygen-of-amplification/.

Poniewozik, James. 2012. PolitiFact, Harry Reid's Pants, and the Limits of Fact-Checking. *Time,* August 8. http://entertainment.time.com/2012/08/08/politifact-harry-reids-pants-and-the-limits-of-fact-checking/. Accessed 27 June 2018.

Rauch, Jonathan. 2018. The Constitution of Knowledge. *National Affairs* 37 (Fall). https://nationalaffairs.com/publications/detail/the-constitution-of-knowledge. Accessed 28 Dec 2018.

Reingold, Howard. 2013. *Crap Detection Mini-Course*, February 20. http://rheingold.com/2013/crap-detection-mini-course/. Accessed 20 Dec 2018.

Schaedel, Sydney. 2017. How to Flag Fake News on Facebook. *The Wire*, July 6 (Updated 28 August 2018). https://www.FactCheck.org/2017/07/flag-fake-news-facebook/. Accessed 4 Jan 2019.

Shapiro, Matt. 2016. Running the Data on PolitiFact Shows Bias Against Conservatives. *The Federalist,* December 16. http://thefederalist.com/2016/12/16/running-data-politifact-shows-bias-conservatives/. Accessed 27 June 2018.

Silverman, Craig. 2007. *Regret the Error: How Media Mistakes Pollute the Press and Imperil Free Speech*, 293–309. New York: Union Square.

Stroh, Sean. 2017. Why the Trusting News Project Is Aiming to Better Understand Public Trust in Journalism. *RJI in the News – Trusting News*, October 10. https://www.rjionline.org/stories/why-the-trusting-news-project-is-aiming-to-better-understand-public-trust-i.

Tripodi, Francesca. 2018. *Searching for Alternative Facts: Analyzing Scriptural Inference in Conservative News Practices*, May 16. https://datasociety.net/output/searching-for-alternative-facts/.

Welch, Matt. 2013. The 'Truth' Hurts: How the Fact-Checking Press Gives the President a Pass. *Reason,* January 20. https://web.archive.org/web/20130120170109/http://reason.com/archives/2013/01/07/the-truth-hurts/1. Accessed 27 June 2018.

Whigham, Nick. 2017. Wikipedia Founder Launches News Website to Combat the Rise of 'Alternative Facts'. *news.com.au*, April 28. https://www.news.com.au/technology/online/wikipedia-launches-news-website-to-combat-the-rise-of-alternative-facts/news-story/905c6bf3d8e02d319ed7ee536e7038f0. Accessed 3 Jan 2019.

Williamson, Kevin. 2015. Politifact and Me. *National Review*, February 26. https://www.nationalreview.com/corner/politifact-and-me-kevin-d-williamson/. Accessed 27 June 2018.

Wolf, Connor D. 2016. RNC: PolitiFact Has a History of Anti-Right Bias. *Inside Sources,* September 26. http://www.insidesources.com/politifact-bias/. Accessed 27 June 2018.

Chapter 6
Where Do We Go Next?

We have taken two topics that are seemingly unrelated, urban legends and political fact-checking, and demonstrated how they are connected to one another and also to other wide-ranging topics such as horror movies, truth-or-fiction television, and computer security efforts. We did this because they serve as a path to a better understanding of how people digest information that is both accurate and misleading, as well as the purposes they use it for. Questions about true and false facts command wide attention across American society, so any insights we can gain from examining them will add considerably to the general understanding of how modern societies use information in many different settings in their everyday lives. The thread that underlies these disparate topics is the act of scrutiny – a fundamental finding of our research. But our discussion led us to conclude that we have barely touched the subject, so this chapter discusses briefly a basic question: What should be explored next regarding the role of scrutiny, particularly as it is affected by the massive use of social media and other forms of information technology?

We observed in Chap. 1 that there are many types of scrutiny: it is an identity feature for certain organizations and professions, e.g. as the methodological basis for science. It is a means for teasing out one's religious or philosophical beliefs, or of understanding the cultural meaning that underlies narratives. It is a foundational precept of honest election and fair governance. It is an underlying characteristic of a practical, rational mindset. It can be exercised for serious purposes or for entertainment purposes. We have seen all of these in the recent world: rampant political fact-checking to protect the integrity of politics, fact-driven government and big business and science, and entertainment such as participation in online discussions of urban legends or watching television programming such as Mythbusters. We have also seen the damage that can be caused when scrutiny is ignored or avoided, such as the lives lost in the Middle East as we sought to destroy the elusive weapons of mass destruction.

The account in this book is intentionally narrowly focused on a few topics. However, the material we have described here is connected to many other narrow and broad topics. Thus, for the remainder of this conclusion, we will focus on six

© Springer Nature Switzerland AG 2019
W. Aspray, J. W. Cortada, *From Urban Legends to Political Fact-Checking*,
History of Computing, https://doi.org/10.1007/978-3-030-22952-8_6

ways in which the scholarship presented in this book can be extended. One of these ways is to broaden the set of scrutinizing activities that have gone on in the United States during the same time period we have covered here, i.e. between 1990 and 2015. One example that we could have covered here is scrutiny by individual consumers of products and services offered in a world replete with advertising. The late nineteenth century witnessed the professionalization of advertising, but it became much more sophisticated throughout the twentieth and twenty-first centuries, and it became more pervasive with the introduction of new technologies such as radio, television, Internet, and interactive social media.[1] The story might include, for example, historical accounts of the Federal Communication Commission, the Federal Trade Commission, the Food and Drug Administration, the Consumer Protection Agency, the Better Business Bureau, and private organizations such as Consumer Watchdog and TINA.org (Truth in Advertising.org).

A second way to build on our research would be to extend the study in space. We have focused on urban legends, political fact-checking, film, television, and computer security issues in the United States. But clearly there are similar phenomena in other countries. For example, we mention briefly in Chap. 3 some urban legends that appeared in the Netherlands and England. Many countries across the globe are experiencing the same anxieties as the United States in dealing with a complex, modern world; and we would expect (but have not studied) that there are efforts elsewhere to come to grips with the associated rumors and legends.[2] Similarly, fake facts are being weaponized by politicians in many different countries, not only in the United States. However, we do not know sufficiently the extent of this misrepresentation nor the extent of the efforts to battle it through the scrutiny of fact-checking.[3] Similarly, there may be foreign films and television programming, or computer security efforts in other countries, that are similar to those that we have covered in the United States.

---

[1] See, for example, Hess (1922); Fox (1984); and Sivulka (2011).

[2] Here is a sampling of the literature available about urban legends outside the United States: Painter (2016–2019); University of Hawaii at Manoa (2017); Hubbard (2018); Ines (2017). Note that none of these sources is scholarly; urban legends from a region are regarded as a form of popular culture. On a more scholarly note, see Bird (2002); Miwa and Mark Ramseyer (2010); Bliesemann de Guevara and Kühn (2015); Simonides (1990); Samels (1998); Top (1990). This is just a sample, but it is clear that, at least written in English, there is a more sizable literature on urban legends in the United States than urban legends in other parts of the world.

[3] According to a 2016 study, 113 fact-checking groups were in existence, spread across 50 countries. See Graves and Cherubini (2016). Japan has been behind other countries in establishing fact-checking organizations (Murakami 2018), but efforts were underway in 2017 to establish a fact-checking coalition (Kajimoto 2017). By 2017, fact-checking had become a major trend in South Korean journalism, in part because of the conflict with North Korea (Boyoung Lim 2017).As of 2018, there were 17 fact-checking organizations in Latin America (Ariel Riera, Fact-checking in Latin America: Features and Challenges, London School of Economics and Political Science, Media Policy Project Blog, March 8, 2018, https://blogs.lse.ac.uk/mediapolicyproject/2018/03/08/fact-checking-in-latin-america-features-and-challenges/, accessed 12 April 2019). This gives a snapshot of fact-checking around the world, but additional information about these activities is available.

A third way to build on our scholarship would be to extend it in time. We have shown in our book, *Fake News Nation*, eight case studies of lying and misrepresentation in American public life, going back 200 years. We have also shown in an article the six exogenous factors that prepared Americans – both in terms of skills and attitudes – for being scrutinizers by the end of the nineteenth century.[4] But it would be useful to examine earlier cases of scrutiny, whether they relate to politics or cultural meaning or other topics from the past, both in the United States and elsewhere. There is much to learn, for example, from the various religious movements competing against each other in the period before the Civil War, how African Americans were characterized and presented both in the nineteenth and twentieth centuries; and that is before we even consider such national controversies as the Vietnam War and the political divides that so separated the similarities of the Democrats and Republicans before the start of the 1960s. A quick reading of nineteenth and twentieth century Western European history casts up a similar collection of topics, made important by the fact that Americans and Europeans influenced each other's thinking and hence how they have approached the act of scrutinizing events and statements.

A fourth way to build on our scholarship would be expand the discussion that we only briefly touched on in Chap. 1, related to authenticity. Authenticity is a kind of truth to self, and the consumers of culture scrutinize cultural artifacts to see if they are authentic and are dissatisfied if they find them wanting on this criterion. The well-known philosopher and literary theorist Walter Benjamin in 1936 published an article entitled "The Work of Art in the Age of Mechanical Reproduction",[5] and this set off a new path of scholarship by art and literary critics, philosophers, and historians such as Anthony Grafton, Denis Dutton, and Arthur Koestler on such issues as the differences between an original painting or piece of music, and a copy or performance; the act of artistic creation and artistic forgery; and the importance of authenticity to the reception of creative works.[6] When we read a novel or watch a play, we question how authentically the author has rendered the setting or the personalities of the individual characters; and this has a significant place in determining our satisfaction with the work. A similar kind of scrutiny in authenticity comes into play when we purchase period furniture reproductions or visit historic Williamsburg. We are convinced that the study of the role of authenticity remains in a primitive state of development, except in a few specific domains such as art. Yet, it needs to be acknowledged as a profoundly important way that people filter and judge the value and veracity of information of many different types. Faith in a source and "gut feel" do not get us to an understanding of its role.

---

[4] James W. Cortada and William Aspray, The Rise of Scrutiny in the United States: Building Capability to Navigate a Complex, Dangerous World in a Time of Lies and Misinformation (paper under review).

[5] The article is reprinted in Benjamin (2008).

[6] See, for example, Grafton (1990); Koestler (1964); and Dutton (1983). Many people have written on this topic. Dutton has an excellent review article: Authenticity in Art, in Levinson (2003).

A fifth way to build on our scholarship is to regard this study as a piece of the emerging field of information history. Both of the authors here have been actively involved in the development of the emerging field of information history. The history of information draws upon previously separate scholarship on the history of books and publishing, the history of libraries, the history of communication, the history of information science, and the history of computing (Aspray 2015). The history of information is deeply intertwined with general history because, at least since the late nineteenth century, information has been recognized as an asset by both businesses and governments; and it has also been seen as important to the everyday lives of people (Cortada 2016a, b, 2019; Aspray and Hayes 2011). The story here connects in particular with the issues of the overabundance of information[7] and the role of misinformation.[8]

This book appears in a series devoted to computer history. So, sixth, how does this book relate to computer history? In a recent article, Jennifer Light argued that the effort to use history of computing to understand the present and the future has been limited by its "device-centered focus" (Light 2019). Light, we believe, would welcome this study because it does not focus on the devices of computing and information technology but instead on the uses of these technologies, and how these uses are tied to political agendas, larger cultural meanings, and other societal forces. As information technology became smaller, more modular and less expensive, it diffused into the hands of the public at large. By the end of the first decade of the twenty-first century, the majority of the American public could gather, read, judge, and then comment on and distribute information over the Internet. The volume of material that flowed in and out of their lives increased by orders of magnitude. Studies of scrutiny and digital literacy enable us to understand better how people work with information. The technology made more information accessible and the features of computing and communications influenced the manner in which people appropriated facts. How all that happened remains a frontier in the study of computing history, just being looked at by scholars.[9]

---

[7] One of the best historical works to appear so far on the overabundance of information is Blair (2011); but also see Levitin (2014); and Rutknowski and Saunders (2019).

[8] On the history of misinformation, see, for example, Anne P. Mintz, *Web of Deception: Misinformation on the Internet* (2002) and *Web of Deceit: Misinformation and Manipulation in the Age of Social Media*; Murray Edelman, *The Politics of Misinformation*; A.P. Napolitano, *Lies the Government Told You: Myth, Power, and Deception in American History*; Denery II (2015); and Hannah Arendt, *Crisis of the Republic: Lying in Politics, ...* (1972)

[9] A secondary contribution of this work to the history of computing is that explores that period just prior to the emergence of the Internet. People were trying to do what they do on the Internet and social media prior to the invention of these technologies, and this study gives insight into some of this pre-history, discussing bulletin boards, Usenet, and private network providers such as America Online.

References

Aspray, William. 2015. The Many Histories of Information. *Information & Culture* 50 (1): 1–23.

Aspray, William, and Barbara Hayes, eds. 2011. *Everyday Information: The Evolution of Information Seeking in America*. London: MIT Press.

Benjamin, Walter. 2008. *The Work of Art in the Age of Technological Reproducibility, and Other Writings on Media*. Cambridge, MA: Harvard University Press.

Bird, S. Elizabeth. 2002. It Makes Sense to Us: Cultural Identity in Local Legends of Place. *Journal of Contemporary Ethnography*, October. https://journals.sagepub.com/doi/abs/10.1177/089124102236541. Accessed 12 Apr 2019.

Blair, Ann. 2011. *Too Much to Know: Managing Scholarly Information before the Modern Age*. New Haven: Yale University Press.

Bliesemann de Guevara, B., and F.P. Kühn. 2015. On Afghan Footbaths and Sacred Cows in Kosovo: Urban Legends of Intervention. *Peacebuilding* 3 (1): 17–35. https://doi.org/10.1080/21647259.2014.969508.

Cortada, James. 2016a. *All the Facts: A History of Information in the United States Since 1850*. Oxford: Oxford University Press.

Cortada, James W. 2016b, August. New Approaches to the History of Information: Ecosystems, Infrastructures, and Graphical Representations of Information. *Library & Information History* 32 (3): 179–202.

———. 2019. Revisiting 'Shaping Information History as an Intellectual Discipline: Shaping Information History as an Intellectual Discipline'. *Information & Culture* 54 (1): 95–126.

Denery, Dallas G., II. 2015. *The Devil Wins: A History of Lying from the Garden of Eden to the Enlightenment*. Princeton: Princeton University Press.

Dutton, Denis. 1983. *The Forger's Art: Forgery and the Philosophy of Art*. Berkeley: University of California Press.

Fox, Stephen. 1984. *The Mirror Makers: A History of American Advertising and Its Creators*, 150–162. New York: William Morrow & Co. 210–216.

Grafton, Anthony. 1990. *Forgers and Critics: Creativity and Duplicity in Western Scholarship*. Princeton: Princeton University Press.

Graves, Lucas, and Federica Cherubini. 2016. *The Rise of Fact-Checking Sites in Europe*. Oxford: Reuters Institute.

Hess, Herbert W. 1922. History and Present Status of the "Truth in Advertising Movement". *The Annals of the American Academy of Political and Social Science* 101 (May): 211–220.

Hubbard, Kirsten. 2018. Central America Folklore and Legends. *Tripsavvy*, updated December 1. https://www.tripsavvy.com/central-america-folklore-and-legends-1490434. Accessed 12 Apr 2019.

Ines, Aaron. 2017. 10 Creepy Urban Legends from The UAE. *Listverse*, August 15. http://listverse.com/2017/08/15/10-creepy-urban-legends-from-the-uae/. Accessed 12 Apr 2019.

Kajimoto, Masato. 2017. A New Fact-checking Coalition Is Organizing in Japan. *Poynter* (June 21). https://www.poynter.org/fact-checking/2017/a-new-fact-checking-coalition-is-launching-in-japan-update/. Accessed 12 Apr 2017.

Koestler, Arthur. 1964. *The Act of Creation*. New York: Macmillan.

Levinson, Jerrold, ed. 2003. *The Oxford Handbook of Aesthetics*. Oxford: Oxford University Press.

Levitin, Daniel J. 2014. *The Organized Mind: Thinking Straight in the Age of Information Overload*, 3–75. New York/Plume: Penguin.

Light, Jennifer S. 2019. Expanding the Usable Past. In *Historical Studies in Computing, Information, and Society: Insights from the Flatiron Lectures*, ed. William Aspray. New York: Springer.

Lim, Boyoung. 2017. What's Behind South Korea's Fact-Checking Boom? Tense Politics, and the Decline of Investigative Journalism. *Poynter* (June 16). https://www.poynter.org/fact-checking/2017/whats-behind-south-koreas-fact-checking-boom-tense-politics-and-the-decline-of-investigative-journalism/. Accessed 12 Apr 2019.

Miwa, Yoshiro, and J. Mark Ramseyer. 2010. *The Fable of the Keiretsu: Urban Legends of the Japanese Economy*. Chicago: University of Chicago Press.

Murakami, Sakura. 2018. Japan Lags in Fact-checking Initiatives, Media Watchers Say. *The Japan Times* (February 16). https://www.japantimes.co.jp/news/2018/02/16/national/japan-lags-fact-checking-initiatives-media-watchers-say/#.XLDacy3MwrU. Accessed 12 Apr 2019.

Painter, Sally. 2016–2019. Urban Legends from Europe. *Love to Know*. https://paranormal.lovetoknow.com/Urban_Legends_from_Europe. Accessed 12 Apr 2019.

Rutknowski, Anne François, and Carol Saunders. 2019. *Emotional and Cognitive Overload: The Dark Side of Information Technology*, 1–16. New York: Routledge.

Samels, Clare A. 1998. Folklore, Food, and National Identity: Urban Legends of Llama Meat in La Paz, Bolivia. *Contemporary Legend*, New Series 1: 21–54.

Simonides, Dorota. 1990. Contemporary Urban Legends in Poland. In *Storytelling in Contemporary Societies*, ed. Lutz Rohrich. Tübingen: Gunter Narr Verlag.

Sivulka, Juliann. 2011. *Soap, Sex, and Cigarettes: A Cultural History of American Advertising*, 325–405. Boston: Cengage Learning.

Top, Stefaan. 1990, January. Modern Legends in the Belgium Oral Tradition. *Fabula* 31 (3): 272–278.

University of Hawaii at Manoa. 2017. *CSEAS, Southeast Asia: Halloween Edition*, October 26. http://www.cseashawaii.org/2017/10/southeast-asia-halloween-edition/. Accessed 12 Apr 2019.

Index

© Springer Nature Switzerland AG 2019
W. Aspray, J. W. Cortada, *From Urban Legends to Political Fact-Checking*,
History of Computing, https://doi.org/10.1007/978-3-030-22952-8